Car Park Designers' Handbook

Car Park Designers' Handbook

Jim Hill
CEng FIStructE (retired)

Glynn Rhodes
BSC (Hons) CEng MICE MIHT

Steve Vollar
EurIng BSc CEng FIStructE MICE

Published by ICE Publishing, One Great George Street, Westminster, London SW1P 3AA.

Full details of ICE Publishing sales representatives and distributors can be found at:
www.icevirtuallibrary.com/info/printbooksales

First published 2005

Other titles by ICE Publishing:

Recommendations for the Inspection, Maintenance and Management of Car Park Structures.
National Steering Committee on the Inspection of Multi-storey Car Parks.
ISBN 978-0-7277-3183-8
ICE Manual of Highway Design and Management. I. Walsh (ed.).
ISBN 978-0-7277-4111-0
Principles of Pavement Engineering, Second edition. N. Thom.
ISBN 978-0-7277-5853-8

www.icevirtuallibrary.com

A catalogue record for this book is available from the British Library

ISBN 978-0-7277-5814-9

© Thomas Telford Limited 2014

ICE Publishing is a division of Thomas Telford Ltd, a wholly-owned subsidiary of the Institution of Civil Engineers (ICE).

All rights, including translation, reserved. Except as permitted by the Copyright, Designs and Patents Act 1988, no part of this publication may be reproduced, stored in a retrieval system or transmitted in any form or by any means, electronic, mechanical, photocopying or otherwise, without the prior written permission of the Publisher, ICE Publishing, One Great George Street, Westminster, London SW1P 3AA.

This book is published on the understanding that the author is solely responsible for the statements made and opinions expressed in it and that its publication does not necessarily imply that such statements and/or opinions are or reflect the views or opinions of the publishers. Whilst every effort has been made to ensure that the statements made and the opinions expressed in this publication provide a safe and accurate guide, no liability or responsibility can be accepted in this respect by the author or publishers.

Cover image: Eastside Birmingham car park, designed by Hill Cannon Consulting LLP and voted Best New Car Park at the British Parking Awards 2012.

Commissioning Editor: Rachel Gerlis
Production Editor: Imran Mirza
Market Specialist: Catherine de Gatacre

Typeset by Academic + Technical, Bristol
Printed and bound in Great Britain by CPI Group (UK) Ltd, Croydon CR0 4YY

Contents

About the authors		ix
Foreword		xi
Preface		xiii
Glossary of terms		xv
Acknowledgements		xvii

01 Introduction — **1**
- 1.1. Historical note — 1
- 1.2. Advice and guidance — 2
- 1.3. Scope — 2
- 1.4. Design flexibility — 2

02 Design brief — **5**
- 2.1. The client — 5
- 2.2. The brief — 5

03 Design elements — **7**
- 3.1. The Standard Design Vehicle (SDV) — 7
- 3.2. Left, right or in the middle? — 9
- 3.3. Parking categories — 9
- 3.4. Aisle widths — 10
- 3.5. Ramps and access-ways — 13
- 3.6. Kerbs — 22
- 3.7. Super-elevation — 24
- 3.8. Parking deck gradients — 24
- 3.9. Headroom and storey heights — 24
- 3.10. Height limitations — 24

04 Dynamic considerations — **27**
- 4.1. Discussion — 27

05 Static considerations — **31**
- 5.1. Static efficiency: discussion — 31

06 Circulation design — **35**
- 6.1. Discussion — 35
- 6.2. How many levels? — 35
- 6.3. Roof considerations — 36
- 6.4. Circulation efficiency — 36
- 6.5. Parking times — 37

07 Circulation layouts — **39**
- 7.1. Discussion — 39
- 7.2. Dimensions used — 39
- 7.3. User-friendly features — 39
- 7.4. Angled and right-angled parking, a comparison — 40
- 7.5. Split-level decks (SLD) — 41
- 7.6. Sloping parking decks (SD) — 53
- 7.7. Flat and sloping deck layouts (FSD) — 67
- 7.8. Combined flat and sloping deck layouts with internal cross ramps (VCM, FSDR and WPD) — 70
- 7.9. Flat parking decks with storey height internal ramps (FIR) — 84
- 7.10. Minimum dimension layouts (MD) — 91
- 7.11. Circular sloping decks (CSD) — 103
- 7.12. Half-external ramps (HER) — 106
- 7.13. External ramps (ER) — 111

08 Stairs and lifts — **121**
- 8.1. Discussion — 121
- 8.2. Vertical and horizontal escape — 122

	8.3. Escape distances	123
	8.4. Lift sizing	124
09	**Disabled drivers and access assistants**	**129**
	9.1. Discussion	129
	9.2. Stall locations	130
	9.3. Stall dimensions	130
	9.4. Access	131
10	**Bicycles and motorcycles**	**133**
	10.1. Discussion	133
	10.2. Bicycle parking	133
	10.3. Motorcycle parking	133
	10.4. Lockers	136
	10.5. Fiscal control	136
11	**Security**	**137**
	11.1. Discussion	137
	11.2. Lighting, music and CCTV	138
	11.3. See and be seen	139
	11.4. Full bay surveillance	139
	11.5. Ground-level enclosure	139
	11.6. Women-only car parks	140
12	**Underground and robotic parking**	**141**
	12.1. Discussion	141
13	**Lighting**	**145**
	13.1. Discussion	145
	13.2. Zoning	146
	13.3. Emergency lighting	146
14	**Signage**	**147**
	14.1. Discussion	147
	14.2. Directional signs	147
	14.3. Information signs	149
	14.4. Variable message sign systems	149
	14.5. Emergency signs	149
15	**Drainage**	**151**
16	**Fire escapes, safety and fire fighting**	**155**
	16.1. Discussion	155
	16.2. Means of escape	156
	16.3. Fire safety	156
	16.4. Fire-fighting measures	157
	16.5. Sprinklers	157
	16.6. Fire escapes	157
17	**Fiscal and barrier control**	**159**
	17.1. Discussion	159
	17.2. Control systems	159
	17.3. Barrier control	161
18	**Ventilation**	**163**
	18.1. Discussion	163
	18.2. Natural ventilation requirements	163
	18.3. Mechanically assisted natural ventilation requirements	163
	18.4. Mechanical ventilation requirements	163
19	**Structure**	**167**
	19.1. Discussion	167
	19.2. Design criteria	168
	19.3. Stability	168
	19.4. Robustness	168

	19.5. Edge protection	168
	19.6. Fire protection	168
	19.7. Vibration	169
	19.8. Durability	169
	19.9. Common forms of structure	169
20	**Appearance**	**173**
	20.1. Discussion	173
	20.2. Appearance requirements	174
21	**Appendix A**	**177**
22	**References**	**179**
	Index	**181**

About the authors

Jim Hill CEng FIStructE (retired)

In 1967, Jim founded the Hill Cannon Partnership with John Cannon and has been involved in car park design since 1969. In 1970, they developed the Tricon structural system and in 1993 Jim patented the Vertical Circulation Module system (VCM). He is a past president of the British Parking Association and a regional chairman of the Concrete Society. As a consultant to the practice, having retired in 1992, he concentrated on the further development of VCM, designing appropriate circulation layouts for many projects and researching this book. He is currently writing a similar handbook on good practice parking in the USA, where he now lives with his American wife Rosie.

Glynn Rhodes BSC (Hons) CEng MICE MIHT

Glynn is Senior Partner of Hill Cannon Consulting LLP and, since 1986, has been involved in the design of 50 multi-storey car parks, two of which have been voted *Best New Car Park* at the annual British Parking Awards. He also received the Ernest Davies Award for the best article published in Parking News, *Current Trends in the Design of Car Parks*. He has provided design advice for large underground car parking facilities in Manila, Kuala Lumpur (Petronas Towers), Zagreb and Dubai. Recent projects include multi-storey car parks at Basildon Hospital, Gatwick Airport, Luton Railway Station, Birmingham International Station, and Nottingham Hub Station, and the one for the Scottish Exhibition and Conference Centre (SECC) in Glasgow.

Steve Vollar EurIng BSc CEng FIStructE MICE

Steve is a Senior Partner of Hill Cannon Consulting LLP and has been actively involved with car park design and parking-related subjects since 1996. He has been responsible for the design of many multi-storey car parks, for inspections, and for the refurbishment of existing structures. His particular interest is in the inclusion of suitably designed disabled parking facilities, general 'wayfinding', and catering for two-wheeled traffic.

Recent projects include Eastside Birmingham which was voted Best New Car Park at the British Parking Awards 2012, Westgate Railway Station Car Park, Wakefield, Southwater, Telford and a 4000-space facility for the Ministries of Finance Complex in Kuwait.

He was chairman of the Yorkshire Branch of the Institute of Structural Engineers and continues to be a professional interviewer for candidates applying for Chartered status.

Foreword

Before retiring as a Consultant to Hill Cannon Consulting Engineers in 2005, Jim Hill had spent over 35 years in the development of car park design and this experience has given him a unique insight into the reasons why some buildings operate successfully and others, of a similar size and activity, do not. The choice of the correct circulation layout is a subject that he considers to be of prime importance in the creation of an efficient parking building.

Both as a consumer of parking services and a former parking manager, it always intrigues me as to why some parking layouts are easily navigated and yet others test one's patience. As an engineer, I think logically and admire the 'art of parking' created by my fellow colleagues; as a consumer I want to be able to park my car as quickly and as effectively as I can and get on with the business in hand, be it work or play, especially if I have children with me.

My experience has taught me that parking is a means to an end; it is the first and last impression of my destination. This is especially true in the retail and commercial world where, hopefully, my custom is valued. Town and city centres rely on good parking facilities, as do many leisure destinations and major transport hubs like train stations and airports. The parking needs to be good if I'm to contemplate returning there again and again.

Equally important is the need to feel safe and welcome wherever I choose to park. Complex layouts, frustration with queues and conflict with others who are manoeuvring in or out of parking spaces, or sometimes in what seems like a never ending set of twists and turns to get in or out of the car park in the first place, only serve to increase my sense of being uncared for by the owner or operator.

This book describes and illustrates some 60+ variations on the many layout themes: their advantages and disadvantages are discussed, recommendations made for their practical application and suggestions made for other layouts that should also be considered.

More than just discussing layouts, the authors have shown how ramps can be prevented from projecting excessively into traffic aisles, how to assess dynamic capacity and efficiency, and the many other considerations that go to make up the design process. The matters dealt with in chapters 8 to 20, such as the current requirements for people with mobility impairments, pedestrian access, security, ventilation, and so on, have been written by Jim's co-authors, all parking experts in their own right.

In Jim's opinion, effective design is based on common sense, a little crystal ball gazing and experience: it is not a precise art. He suggests that if drivers want to use a car park and clients are willing to pay for it, little else matters. I wouldn't want to disagree with him, but I believe my comments about feeling welcome at any parking facility are the key to its success. If the operator wants to do business, good customer service is vital, and this depends on good design.

This second edition of the book updates and expands on the subjects covered by the first edition in relation to all aspects of car park design, especially the design of circulation layouts, in a practical manner, and can be easily understood by anyone with an interest in the subject. It will help to identify examples of best practice in making our parking facilities more accessible to all. The book is a useful reference for those considering Park Mark® – The Safer Parking Scheme (police-approved safer parking).

Kelvin Reynolds
Director of Policy and Public Affairs at British Parking Association

Preface

Information on the design of vehicle circulation systems in car parks is hard to find: had it not been so this book, probably, would not have been written. To our knowledge, special features and relative efficiencies of car parks have never before been discussed in any great detail. Many designers are unaware of the advantages of using a particular layout system over another and it is a major purpose of this book to redress that imbalance.

No matter how efficient the structural solution and how attractive the architectural appearance, if it is wrapped around a poor choice of circulation layout the result will be another unpopular car park. In many under-used car parks the reason for their unpopularity is not that they have been allowed to become dirty and/or dingy (conditions that by themselves would not normally put off motorists), but rather that they suffered from a poor choice of internal layout. Of the many buildings inspected, the most unpopular have, invariably, incorporated inappropriate circulation designs. Rather than giving these car parks an expensive cosmetic makeover, the money would have been better spent on improving the layout, even at the cost of losing a few parking stalls.

Over the years, as the authors became more experienced, so their awareness of the number of different layouts available increased. Some have been rejected as being impractical or just plain whimsical, but those that are featured in this book are practical and have been constructed somewhere, although not always in the UK. With more than 7000 car parks in the UK, 35 000 in the USA and many thousands more in the rest of the world, it is unlikely that all of the possible variations will have been covered and if any reader is aware of a practical circulation layout substantially different from those featured and lets us know, if it is included in a future edition they will be acknowledged as the source.

The design of underground car parks has also been addressed. In general terms, it follows the recommendations for parking above ground and differs only in the requirements for ventilation. Central points for these layouts can be located either above or below ground. It will be necessary to discuss the proposed surface treatment (large trees and bushes) with the landscape architect before reaching a final decision on the circulation layout and for this reason features above ground have not been shown.

The design of a standard car park is an engineering discipline that requires specialist knowledge in all aspects of car park design, management and operations. Desirable elements in car park design include

- well-laid-out parking bays, aisles and ramps to ensure ease of circulation for both vehicles and pedestrians
- user-friendly environment – with good ventilation, lighting, and open areas to ensure good natural surveillance
- economy in achieving good gross floor area per space ratios
- adequate access and gross control to ensure good dynamic traffic-flow characteristics
- economical structure that is easy to build and is economical while providing long-term durability.

It is the authors' intention that the publication of this handbook will assist the designer in achieving their objectives.

Glossary of terms

Access-way or cross-way	A traffic lane without adjoining stalls, laid flat or to a slope not exceeding 5%; also capable of being used by pedestrians
Aisle	A traffic lane with adjoining stalls on one or more sides
Bin	The dimension across an aisle and its adjacent stalls (a half bin has stalls only on one side)
Circulation efficiency	A comparison (given as a percentage) of the travel distance required to search the stalls, in any particular car park, with the minimum travel distance
Congestion	Applies to traffic that is unable to flow freely
Cross-ramp	An inclined traffic lane connecting the aisles in adjacent bins, laid to a slope greater than 5%
Deck	A single floor that extends over the plan area of a parking building
Des Recs	A shortened form of words describing the *Design Recommendations for Multi-storey and Underground Car Parks, 3rd edition*, published in June 2002 by the Institution of Structural Engineers
Dynamic capacity	A measure of the rate that traffic can pass a given location within a car park (given in vehicles per hour)
Dynamic efficiency	A measure of the ability of a car park to process vehicles under normal operating conditions
Excluded	Applies to an inflow route that is separated from an outflow route
Extended	Applies to any traffic route that is not rapid
Included	A flow route that is located within the circulation pattern of another
Inflow	Applies to the search path for traffic within a car park
Manoeuvring envelope (ME)	The boundaries established by the minimum turning circle when entering a cross-way or ramp, outside of which a vehicle is unable to manoeuvre without reversing
MPV	Acronym for a multi-purpose vehicle
MSCP	Acronym for a multi-storey car park
One-way flow	Traffic flowing in a single direction on an aisle
Outflow	Applies to traffic exiting from a car park
Ramp	Any traffic lane, without adjoining stalls, that provides access to or from parking at different levels
Rapid	Applies to a short route for inflow or outflow traffic
Stall	The parking area allotted to a single vehicle, exclusive of any other adjoining area
Stall pitch	The spacing for stalls, normal to an aisle, for a particular angle of parking
Static capacity	The total number of stalls contained within a designated area or complete car park
Static efficiency	The area of the parking decks divided by the static capacity and given as an area per stall
SUV	Acronym for a sports utility vehicle
Swept path	The width on plan established by a vehicle for any given radius of turn
Two-way flow	Traffic flowing in both directions on an aisle, ramp or cross-way
Vph	Vehicles per hour

Acknowledgements

All photographs featured are from the archives of Hill Cannon Consulting, except

Figure 10.3 courtesy of Falco UK Ltd

Figure 10.4 courtesy of Motoloc Ltd

Figures 19.2, 19.3, 19.4, 19.5 and 19.6 courtesy of Ramboll.

A special acknowledgement to Nigel Dobson, CAD specialist for Hill Cannon, for his assistance, advice and help in the production of the 30 drawings.

Car Park Designers' Handbook
ISBN 978-0-7277-5814-9

ICE Publishing: All rights reserved
http://dx.doi.org/10.1680/cpdh.58149.001

Chapter 1
Introduction

Rules and regulations are but the paper bastions behind which the inexperienced fight their battles: in the matter of car park design let common sense prevail.

1.1. Historical note

Eugène Freyssinet, 1879–1962, a French structural engineer and the inventor of pre-stressed concrete, is credited with designing the first European multi-storey car park in 1920, a split-level layout (SLD 1 type). In the UK, the first multi-storey car park was built *c*. 1924 and it is conservatively estimated that there are now well over 7000, many of them constructed in the post-war boom years between 1950 and 1975.

In the early years, the little information that was available concerning the design of this new type of building was mainly to be found in technical literature distributed by specialist car parking firms in the construction industry. The manoeuvring geometry of vehicles, however, imposed a strong practical discipline resulting in the general principles of layout and design rapidly becoming rationalised, with split-level decks and one-way traffic flows used in many of the earliest buildings designed specifically for parking cars.

Independent information gradually became more available, especially after the publication in 1969 of the *AJ Metric Handbook* and in 1970 of *BPA Technical Note 1. Metric dimensions for car parks – 90° parking*. In 1973, at a Joint Conference on Multi-storey and Underground Car Parks, organised by The Institution of Structural Engineers and The Institution of Highway Engineers, a paper by BR Osbourne and WP Winston was presented containing most of the relevant information available at that time relating to parking geometry. The 'Des Recs' (Design recommendations for multi-storey and underground car parks) were published in 1976: the first attempt to create a national standard work on car park design. It contained much of the information presented at the Joint Conference, together with relevant parts of *Report LR 221* by the Road Research Laboratory, *Parking: Dynamic Capacities of Car Parks*, published in 1969. The report was omitted from subsequent editions.

Historically, MSCPs in the UK have suffered an unenviable reputation for poor layout design and quality of parking. The problem has always been in balancing the motorist's desire for ample room in which to park with the client's desire to build as economically as possible, but in a highly competitive market designers sometimes went too far in the direction of pure economy, and cost-conscious clients were insufficiently critical about poor design features. This resulted in car parks that were lacking in essential dimensions. Many were poorly constructed, badly lit, inadequately waterproofed, awkward to park in, and insecure: mostly, however, they had the merit of being cheap to build.

Modern social trends recognise that parking quality plays a major role in the choice of destination for motorists and their families. Car parks provide, quite often, the first and last impression that people experience when visiting an urban location or commercial enterprise, and have a significant influence on any decision they may make to return. Over the years there has been little change in the manoeuvring envelope of cars licensed to drive on the public highway. Figure 1.1 shows that even in 1910 car-manoeuvring envelopes were similar to those of today's vehicles and that a well designed modern car park can be used by vehicles of all ages.

Figure 1.1 A 1910 Ford in a 1990 car park

1.2. Advice and guidance

Multi-storey car parks are utilitarian constructions. Their design is not a finite art: it is a compromise, a balancing act between motorists' spatial desires and the practical need to achieve economy of construction and effective use of the site area. Stall dimensions, aisle, ramp, and access-way widths, ramp slopes, headroom and circulation layouts can all vary: the critical criterion is general acceptance by the motorist. The purpose of this handbook is to provide advice and guidance on the features that will enable car parks to perform their function efficiently and economically and to be, at the same time, 'user friendly'.

1.3. Scope

The contents of this handbook cover the practical aspects of design for self-parking facilities. Block parking and valet parking, where attendants park cars, have been excluded. Also excluded are mechanically operated car parks and matters concerning architecture, except where they are affected by practical considerations.

1.4. Design flexibility

A multi-storey car park, whether above or below the ground, is costly to construct. Consideration should be given to possible changes of parking function during its working life. Initially, it might seem sensible and economical to provide minimum dimensions and standards to suit a particular purpose. Within time, however, its parking category may change, and unless the interior layout is sufficiently flexible to cope with these changes, the facility could become redundant.

Example 1 is a multi-storey car park serving a town centre bus station, which was relocated. The site was sold for retail development and the car park was offered as part of the deal. The layout with warped decks, although adequate for its original purpose, was not suitable for shoppers and so a 500-space building in good order had to be demolished.

Example 2 is a large factory, located on the edge of town, which closed down and was sold for retail development. The car park for the workforce, designed to minimum standards, was unsuitable for use by the general public. It too, had to be demolished.

Introduction

In the early 1970s, it was rarely considered that car parks could be bought and sold, and that the role for which they were originally designed might alter. Nowadays, however, they are being bought and sold in increasing numbers, either individually or collectively. Market values depend not so much on their architectural merits but on their popularity with the parking public and as such they should be designed, within reason, to be as flexible as possible.

Chapter 2
Design brief

2.1. The client

Not all clients have an expert knowledge about car parks. They are conscious of the need to provide a certain number of parking spaces at a given location, but they are not, necessarily, aware of the information that is required in order to produce the most efficient and cost-effective building.

Designers should present their clients with a questionnaire in order to obtain the maximum amount of relevant information as early as possible. It is unlikely that it will all be available at the preliminary stage, but it does no harm to ask the questions and, at least, it establishes the designer's expertise in the subject.

2.2. The brief

Apart from items such as a ground investigation report and an accurate site and level survey, both of which may require an unacceptable financial outlay by the client at an early stage, the brief should include as much of the following information as possible.

- The maximum and minimum number of spaces required.
- A plan of the proposed site to a known scale, showing the building lines and the surrounding access roads.
- Site levels, even if they are only approximate.
- Proposals for future development that might have an effect on the setting out, shape and function of the building.
- The presence and if possible, the location of electric cables, gas pipes, drains and sewers that might be under the site, especially those that must not be moved or built over without special precautions being taken.
- The maximum number of parking levels and height of building required by the client or allowed by the local planning authority.
- The proposed use, whether it is to be a long-, medium- or short-stay facility, together with the anticipated vehicle entry and exit traffic flow figures.
- The category of parking required, bearing in mind the possibility that it could be sold at some future date for another purpose category.
- The proposed method of payment to be adopted for financial control, whether it is to be a 'payment on exit', 'payment on foot', or 'pay and display' system, or even no payment at all.
- The client's preference, if any, for a particular type of construction.
- The accommodation required for staff and the general public (offices, rest rooms and toilets, etc.).
- The capacity and preferred location for lifts and/or escalators. This is an especially important item when in conjunction with retail shopping.
- The requirements for water-protection over the top parking deck, either with asphalt or an elastomeric membrane, or leaving the top deck untreated, or even roofing over the complete building.
- The need to provide heating to exposed ramps.
- Whether or not a mechanical means of ventilation is acceptable: an important issue when the building is close to adjacent site boundaries.
- The client's preference, if any, for standards of finish in lift lobbies, escape stairs, on parking floors and on exposed parts of the internal structure
- The levels of interior lighting to be adopted (if it varies from British Standards) and any other instructions regarding painting of the interior that could affect the lighting design.

- The requirements for security, CCTV, patrols on foot, or any special emphasis on 'user friendly' aspects.
- The standards required for external and internal signs (illuminated and painted).

The provision of water and power supplies for cleaning purposes.

Car Park Designers' Handbook
ISBN 978-0-7277-5814-9

ICE Publishing: All rights reserved
http://dx.doi.org/10.1680/cpdh.58149.007

Chapter 3
Design elements

3.1. The Standard Design Vehicle (SDV)

3.1.1 Discussion

More than 50 different car manufacturers offer some 340 basic models for sale to the general public in the UK, and model variations increase the choice to over 400. Add to that the makes and models that have been discontinued over, say, the past 15 years but can still be seen in reasonable quantities, and the number rises to well over 500. In size, they range from the diminutive smart car up to the stretch limo.

To build a car park that can cater for them all is unrealistic and uneconomic. Large, limousine-type vehicles occur in relatively small numbers and very few are ever likely to require parking within a structured public car park. It has, therefore, become established practice to design car parks to readily accept the smallest 95% of privately licensed vehicles registered to drive on the highway (see Appendix A). This is not to imply that some of the larger vehicles must not enter a parking building, but they must do so with greater care than the 95-percentile vehicle.

The introduction of the multi-purpose vehicle (MPV) and the proliferation of four-wheel drive (4WD) vehicles in private ownership have resulted in an increasing use of bulkier and taller vehicles by the motoring public. It is likely that the trend will be towards even greater numbers of this type being parked, as leisure activities expand and become an increasingly important factor in the choice of personal transportation. Although not frequent visitors, it would be an advantage for larger vehicles to be able to circulate, even if they have to overflow into adjacent stalls in order to park. The most testing time can sometimes be when a building is officially opened and dignitaries or CEOs in their official limousines are taken on a ceremonial drive through the car park (it has happened).

3.1.2 Length and width

A rectangle 4.8 m by 2 m on plan will accommodate 95% of privately owned vehicles in the UK. The width is measured to include the wing mirrors. Without wing mirrors, the width can be assumed to be 1.8 m (see Figure 3.1).

3.1.3 Height

In height, most cars are less than 1.5 m, but there are a growing number of 4WD and sport utility vehicles (SUV) using car parks, that should not be ignored. Of these, among the tallest currently in use, without roof racks, are the Land Rover Defender (2.035 m) and the Discovery (1.919 m). While they have never been sold in large numbers, they can still be frequent visitors, especially to provincial and market town car parks. Vehicles made for volume sales will, most probably, always be capable of being driven into a standard domestic garage and it is unlikely that they will ever exceed a height of 1.95 m. Camper vans, also, do not always fall within the SDV envelope, but in some resort car parks there may be a need to accommodate them, even if only at the ground parking level.

3.1.4 Wheelbase

A wheelbase of 2.9 m is used to provide the worst-case scenario for changes of level at steep ramps and inclines (see Figure 3.1).

3.1.5 Ground clearance

Although the normal ground clearance for the SDV is better than 120 mm, a well-laden vehicle could be less, especially at the rear end. A dimension of 100 mm, therefore, is considered to be a reasonable minimum for design purposes (see Figure 3.1).

3.1.6 Turning dimensions

Vehicles operating within the SDV envelope are capable of turning between wall faces 12 m apart (see Figure 3.1). Many large vehicles can also turn within this diameter but a very few

Figure 3.1 The standard design vehicle (SDV) (a composite of 95% of private vehicles registered to drive on the highways)

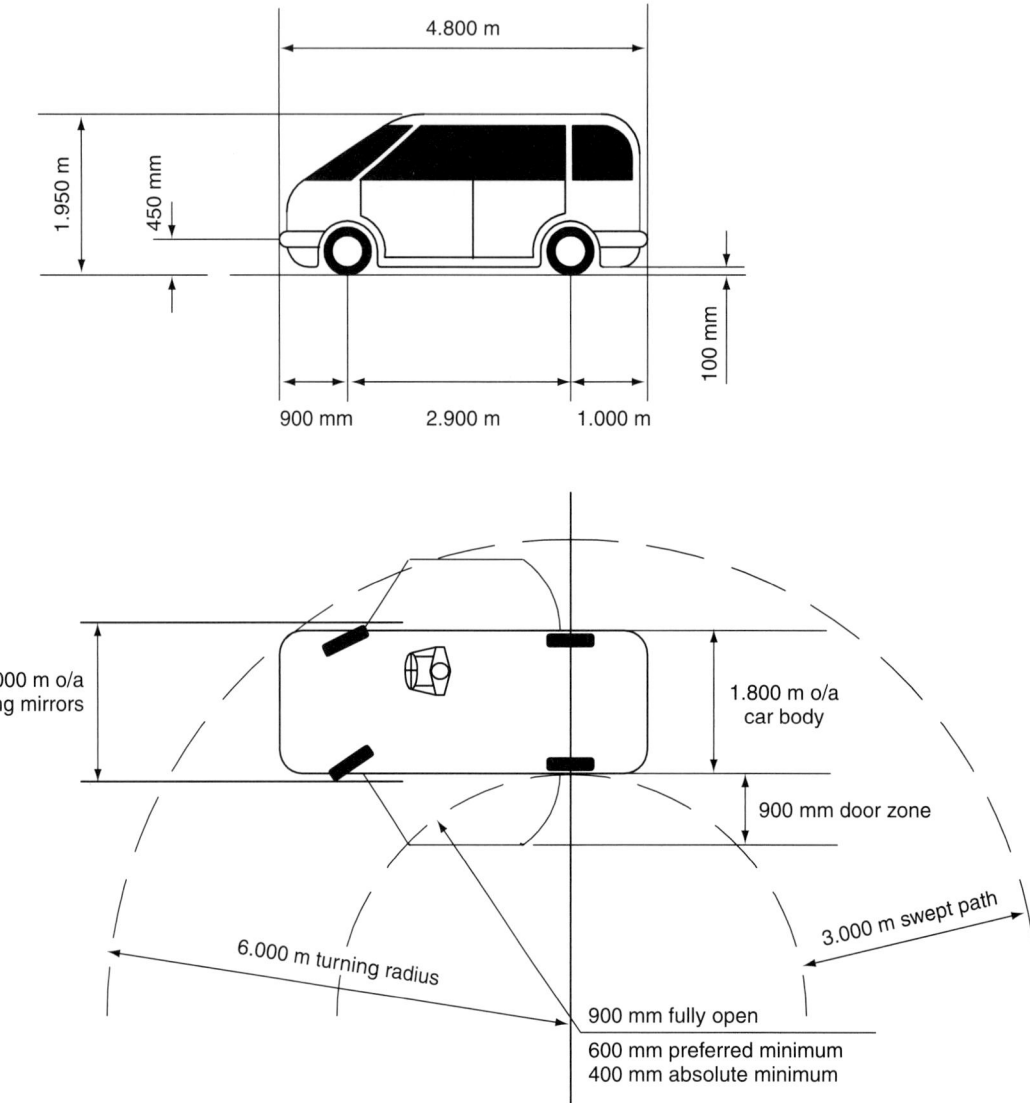

need as much as 15 m to complete an 180°, wall to wall, turn on full lock. At a speed of 10 mph (16 kph), it takes approximately 4.5 m (one second's driving) to develop a 90° turn with a radius of 9 m, and this has to be taken into account when considering the turning manoeuvre.

It is unreasonable to expect motorists to drive around a car park to the extreme manoeuvring ability of their vehicle. Long before this condition is reached, they will have abandoned the building for less onerous places to park, but occasionally it will be necessary to use a full lock turn when entering a stall, or to avoid a pedestrian or another vehicle. This is acceptable but, for good parking practice, motorists must be given the ability to manoeuvre readily in either direction. As a general rule, the minimum turning circle for manoeuvring between adjacent aisles should be in the order of 150% of the SDV turning circle where 90° turns into and out of cross-aisles and ramps are the norm, and 200% where 360° turns are anticipated.

3.1.7 Recommended minimum diameters for turns up to 180° between obstructions

For good practice	18 m
Tight, for 'long-stay' with light usage	16 m
Very tight, for private use only, on small awkward sites and with the clients' prior agreement	14 m
Entering and leaving parking stalls	12 m

Design elements

3.2. Left, right or in the middle?

When vehicles are being driven along a one-way flow aisle it has been observed that they tend to keep towards the centre and so, for right-hand-drive cars, drivers will be biased to the right-hand side of the aisle. This provides them with a better viewing angle to observe the status of stall openings and cross-ways on the left than those on the right. For two-way flow aisles the situation is reversed. The shallower the parking angle becomes, the less significant this becomes and at angles less than, say, 60°, the condition does not occur. It is not a major factor but it is useful to know that stalls and cross-ways on the left are more appreciated in one-way flow facilities, and the opposite situation occurs in two-way flow facilities.

3.3. Parking categories
3.3.1 Discussion

The correct choice of circulation layout and stall dimensions to suit a particular building purpose can be an important factor in the success or failure of a parking facility. For main transportation terminals, it is unlikely that the category will alter, and car parks can be designed with confidence. For most other 'town centre' types, however, changes can and do occur, and this possibility should be considered at the design stage. Four categories of parking are described as follows.

3.3.2 Parking categories

Category (Cat.) 1	Short-stay, for intensive usage with high turnaround rates, usually associated with busy supermarket-type shopping activities
Cat. 2	Medium-stay. Urban centre-type car parks for mixed business, visitor and town centre shopping
Cat. 3	Long-stay. Located at major transport terminals where the flow is intermittent and mainly light, but continuous. Short periods of intensive vehicle movement can also occur when a large people transporter disgorges its passengers
Cat. 4	Tidal, such as occurs in staff car parks where the traffic flow is inwards in the mornings and outwards in the evenings.

3.3.3 Parking stalls
3.3.3.1 Discussion

It is has become normal practice in the UK for designers to adopt stall widths of between 2.3 m and 2.5 m, depending on the parking category. For specific purposes, this can vary, but it must be appreciated that stall widths are an important factor, affecting both flexibility and market values.

The prime consideration is not so much the overall width of a stall but the gap between parked vehicles. Altering the pitch by 100 mm only has a 4% effect on the stall width, but it can result in a 20% variation on the gap between cars and be the difference between getting out of the car with ease, or with some difficulty. 600 mm is considered to be the minimum space that enables most drivers to access their vehicles.

Most vehicles are narrower than the SDV, and so for a stall width of 2.4 m the gap between cars will usually be greater than 600 mm. It is also the case that some drivers are not particularly mobile, while others can be rather large and need a greater door-opening distance.

In Figure 3.1, it can be seen that the full door-opening width is about 900 mm, resulting in an optimum stall width of 2.7 m. Averaging between large and small vehicles, a 2.6-m-wide stall could also produce a gap of about 900 mm, but then economic factors come into play. The compromise answer is shown in Section 3.3.4 and has been generally recognised over many years as an acceptable balance between space and cost.

3.3.4 Recommended dimensions for differing parking categories

All stall lengths	4.8 m
Minimum stall widths	
Cat. 1 (Less than three hours' stay per car)	2.5 m
Cat. 2 (More than three hours' stay per car)	2.4 m
Cat. 3 (More than 12 hours' stay per car).	2.3 m
Cat. 4 (Staff type, mainly tidal)	2.3 m
Disabled drivers	3.6 m
Assistants to disabled drivers	3.2 m

Figure 3.2 Columns located between adjacent stalls

It should be appreciated that the market value of the building could well be affected by the choice of stall width, and stall widths of less than 2.3 m cannot be recommended for general public use. In specific locations, such as hotel or staff-type parking, stall widths of 2.2 m and even 2.1 m have been used where there is a desperate need, where smaller cars are the norm and the client is fully aware of the reduction in parking standards.

3.3.5 Obstructions between stalls

The standard stall widths assume that there are no obstructions between adjacent stalls and that car doors can open freely into the spaces between parked vehicles. It also assumes that drivers and passengers can pass between adjacent cars to gain access to the traffic aisle. If obstructions such as structural columns occur between stalls, the recommended widths should be capable of being measured between the column faces, at the very least. Where rows of more than six stalls occur between wall or column faces, it is not usually necessary to increase the end stall widths as a high proportion of cars are smaller than the SDV and they can park without difficulty at these relatively few locations (see Figure 3.2).

3.3.6 Angled parking

Angled stalls ease the parking manoeuvre. The shallower the angle, the easier and simpler it is to park. Their use is generally restricted to one-way traffic flows, and as the parking angle reduces so can the aisle width necessary for manoeuvring in and out of the stalls: in the process, however, the people/vehicle separation distance is reduced and the floor area per stall requirement is increased (see Figure 3.3).

3.4. Aisle widths
3.4.1 Discussion

Aisle widths can vary according to the traffic flow pattern and the parking angle. The dimensions shown in Section 3.4.4 are adequate for manoeuvring into and out of parking stalls, but no allowance has been made for pedestrians mingling with car traffic on the aisles.

With 90° parking, a 6-m-wide traffic aisle enables pedestrians to walk down a 2-m-wide lane on each side of a centrally located vehicle; alternatively, pedestrians can walk down the central part of the aisle and cars can drive by on either side.

When 45° parking is adopted, the aisle width can be reduced to 3.6 m; however the space available to pedestrians is thus reduced to 800 mm on each side of a centrally located vehicle, and the turning dimension between bins is less than the minimum recommended dimension. In such cases, designers should consider whether some upward dimensional adjustment is desirable, especially in Cats 1 and 2-type facilities where the minimum recommended aisle width is 5 m.

Design elements

Figure 3.3 Comparison of the deck area per stall for three angles of parking (excluding ramps and access-ways)

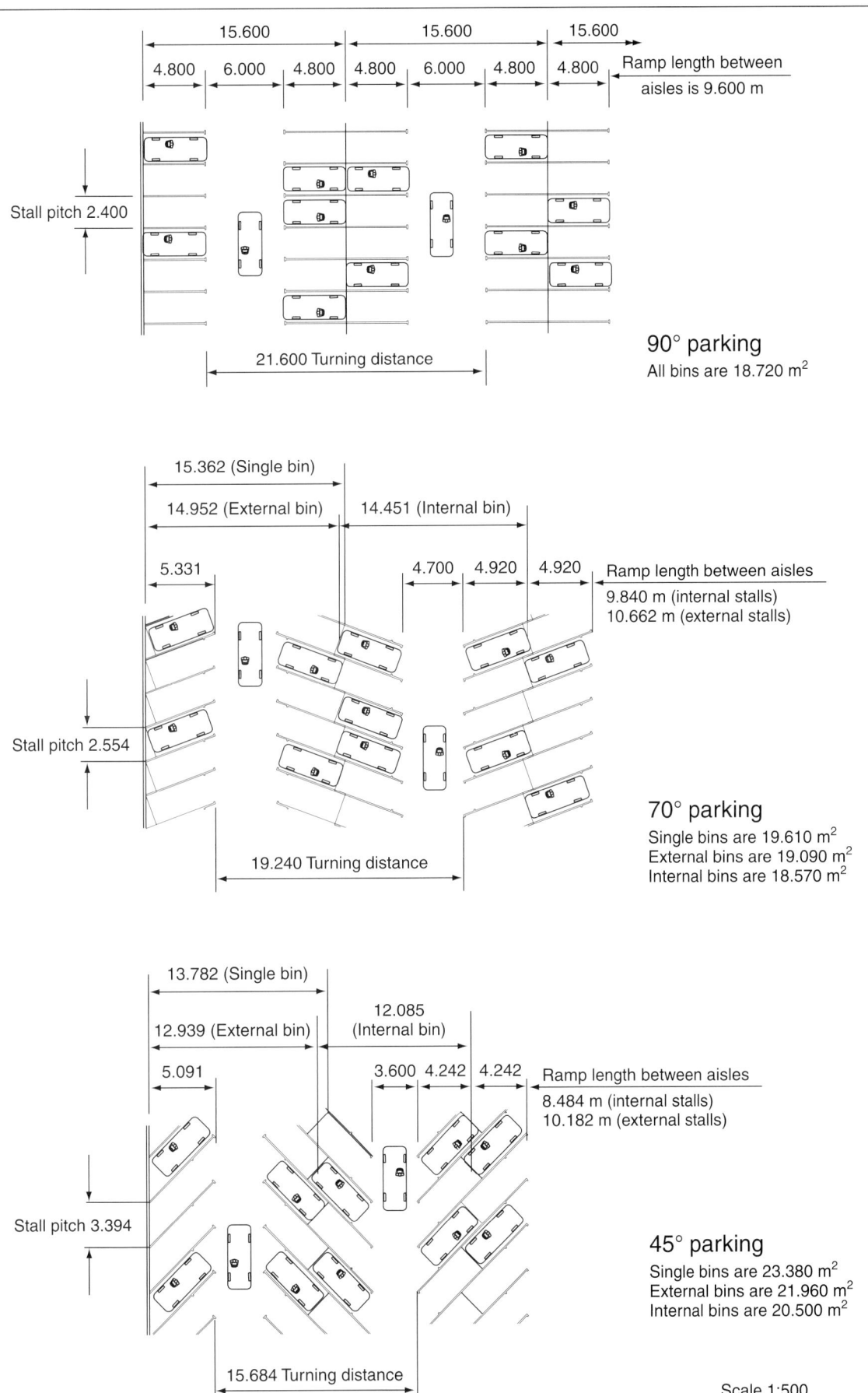

Car Park Designers' Handbook

Surface parking, incorporating 60° parking angles with widened, two-way flow aisles, has been noted in some states in the USA. In such cases, the stalls have been angled so that parking can only, realistically, be achieved on one side at a time. The stall search pattern is greatly extended and the only advantage appears to be in increasing the separation distance between vehicles and pedestrians on the aisles.

3.4.2 One-way flow with reduced aisle widths

Figures 3.10(a) and 3.10(b) show the entry envelope for 2.4-m-wide stalls. It can be seen that an aisle width of about 6 m is required for 'straight in' parking. Increasing the stall width enables drivers to manoeuvre in and out more easily and can result in a reduction in the width of the aisle without reducing the original parking standards. Figures 3.10(c) and 3.10(d) show the reduced aisle widths that can be used for 2.5- and 2.6-m-wide stalls, although such reductions should only be used when site conditions dictate the situation.

3.4.3 Two-way flow with reduced aisle widths

When the anticipated traffic flow is 'tidal', such as occurs in facilities dedicated to a factory work force and/or office staff, two-way flow layouts have been used successfully with aisle widths similar to those recommended for one-way flow circulation. Consideration, however, should be given to the possibility of future changes in parking use that could affect the continued effectiveness of the building.

3.4.4 Manoeuvring on aisles

Recommended minimum aisle widths

90° with two-way flow	7.00 m
90° with one-way flow	6.00 m
80°	5.25 m
75°	5.00 m
60°	4.20 m
45°	3.60 m
90° with one-way flow	
2.5-m-wide stalls	5.70 m
2.6-m-wide stalls	5.40 m

It is not recommended that reduced aisle widths should be adopted as a general rule, but in extreme situations some dimensional flexibility is available to the designer. Where intensive use by pedestrians is anticipated (Cats 1 and 2), aisle widths of less than 5 m cannot be recommended, regardless of the parking angle.

3.4.5 Turning between aisles

A factor to be considered is that as the parking angle decreases so, also, does the dimension available for turning between adjacent aisles. For 90° parking, the clear turning dimension for two traffic aisles and the pair of stalls between them is 21.6 m, but at 45° it reduces to 15.684 m and is below the recommended minimum turning diameter (see Figure 3.3).

3.4.6 Bin dimension
3.4.6.1 Discussion

Bin widths are the sum of the aisle width and the adjacent stalls measured normal to the aisle. With angled parking this dimension will vary, depending on the width of stall chosen. Where multi-span flat decks incorporate angled parking, bins on the external rows will differ in width from those on the internal rows due to the interlocking effect of the stalls. They will also differ from those bin widths generated by a single parking deck (see Figure 3.3).

3.4.7 Recommended minimum bin dimensions for parking with 2.4-m-wide stalls (in metres)

Angle	Single	External	Internal
90°	15.600	15.600	15.600
75°	15.515	15.200	14.894
60°	14.914	14.314	13.714
45°	13.782	12.939	12.085

For two-way flow, the only logical angle is 90° and the recommended minimum bin width is 16.6 m.

Figure 3.4 Shear wall has been set back from the adjacent aisles and openings provided to maximise visibility

3.5. Ramps and access-ways
3.5.1 Discussion

Cross-ramps and access-ways linking adjacent parking decks are one of the most important elements governing driver appreciation. If they are too narrow or too steep, motorists will shun the car park. The entrance should be of a width that will enable drivers of average ability to enter at 10 mph from the optimum position on an aisle without the need to make fine judgements on driving accuracy. In the middle section and where they exit into a wider traffic aisle, the ramp can be reduced in width. After incorporation into a building, ramps are extremely difficult to move or alter (see Figures 3.4–3.6).

Under normal operational conditions it will, occasionally, be necessary for drivers to avoid other traffic and/or pedestrians and commence their turn from other than the optimum

Figure 3.5 Vehicle scrape marks on the outside wall of a 3-m-wide ramp

Figure 3.6 A pair of 4.4-m-wide open aspect ramps

position on an aisle. In such cases, it should be possible to tighten the turning circle and still enter the ramp in one manoeuvre. The wider the ramp entry, the more flexible the aisle position can be. It is unrealistic and uneconomic to attempt to cater for the worst-case situation, but it is equally unrealistic to assume that the optimum location on an aisle will always be available. A clearance dimension of 300 mm should be incorporated on the outer side of aisles when establishing the turning dimension.

It should be noted that at the bottom exit from a ramp, the effect of a vehicle straddling the change of slope increases its vertical height above the deck. It is particularly important to check the headroom at this location.

Modern vehicles should be able to negotiate inclines of up to 25% without difficulty and many such slopes occur on highways the world over. However, it is not just the incline that governs the limiting factors for car parks but the proximity of pedestrians and their safety, coupled with the daunting appearance that a steeply inclined ramp in an enclosed space presents to motorists.

Recommended vehicle ramp gradients are not the result of precise calculation, but are generally considered to be those that most motorists will accept without undue resistance. Small variations to overcome local problems can be tolerated, but as a general rule, designers should be wary of increasing the recommended figures by any significant margin.

The regulations for pedestrian ramps are based on BS 8300 (BSI, 2010) relating to places of work and residence wherein disabled persons can spend many hours per day. In such places it is important to address issues of accessibility for people with restricted mobility; however, most people – regardless of disability – spend only a few minutes at a time in a car park.

Slopes of 8% that raise half a storey, once, on the way to a lift and the building exit (see also chapter 9) are navigable by most people. In many instances, the street outside a car park will slope far more than British Standards allow within buildings (see Figure 3.7).

3.5.2 Recommended maximum vehicle gradients

Straight and helical vehicle ramps
Up to a maximum ramp rise of 1.5 m the steepest sloping element should not exceed 17% (about 1 in 6), reducing at the rate of 1% for each 220 mm increase in ramp height up to 3 m. Above a ramp height of 3 m, the maximum sloping element should not exceed 10%.

Design elements

Figure 3.7 Hillside conditions appear in many car parks

Pedestrian ramps
Less than 2 m going   8.5%
2 m to 5 m going   6.6%
5 m to 10 m going   5.0%

3.5.3 Transitional slopes

It can be demonstrated (see Figure 3.1) that when the minimum ground clearance of the SDV is 100 mm a transition length of 900 mm is all that is necessary to master a ramp slope change of 19% (1 in 6 plus 1 in 50). They also smooth out the change from a ramp slope to a flat deck, but it can be argued that sudden slope changes help to deter speeding within a car park. When long transition slopes are used, the ramps can project well into the traffic aisles (see Figure 3.8(a)), creating obstructions to pedestrians. Restricting their length to the minimum acceptable to motorists is a desirable condition (see Figures 3.8(b) and 3.8(c)). Transitional slopes are not required when the ramp slope change is 10% or less.

3.5.4 Ramp projections into aisles

Figure 3.8(a) shows a ramp rising 1.5 m that agrees with clause 4.3.8 of the 'Des recs'. Transition slopes each with a length of 3 m result in a 1.2 m projection into the aisles. At the aisle side the resulting wedge will be 100 mm high and will need to be feathered out across the entry into the adjacent stalls. Where scissors-type ramps occur, the step between adjacent ramps will be 200 mm. As the storey height reduces, so can the projection into the aisles and when the storey height reduces to 2.4 m the projecting wedges can be eliminated.

The flat decks on either end of the ramps should incorporate drainage falls of 2% (1 in 50). Figure 3.8(b) shows the ramp shape required when they fall towards the ramp and the aisle projection is restricted to 600 mm. The transitional slope becomes 2.9 m at the bottom of the ramp and 1.2 m at the top. Although this is different to the 'Des recs' recommendations, it is a perfectly acceptable alternative reducing the step height to 43 mm. When the storey height becomes less than 2.7 m the aisle side projection can be eliminated.

When the flat decks fall away from the ramp, Figure 3.8(c) shows the ramp shape where at the bottom, a maximum slope change of 10% occurs. This option enables the ramp projection to be 300 mm and omitted altogether when the storey height is less than 2.85 m.

Figure 3.8 Vehicle ramps with 1.5 m half-storey height

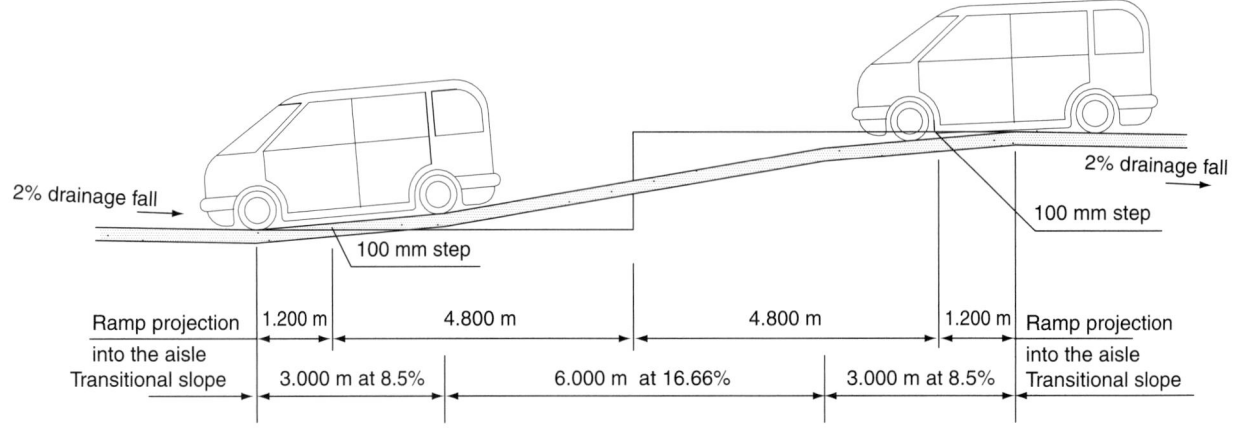

(a) Ramp conforming to the Des Recs, clause 4.3.7
(Ramp projections do not occur when storey heights are less than 2.600 m)

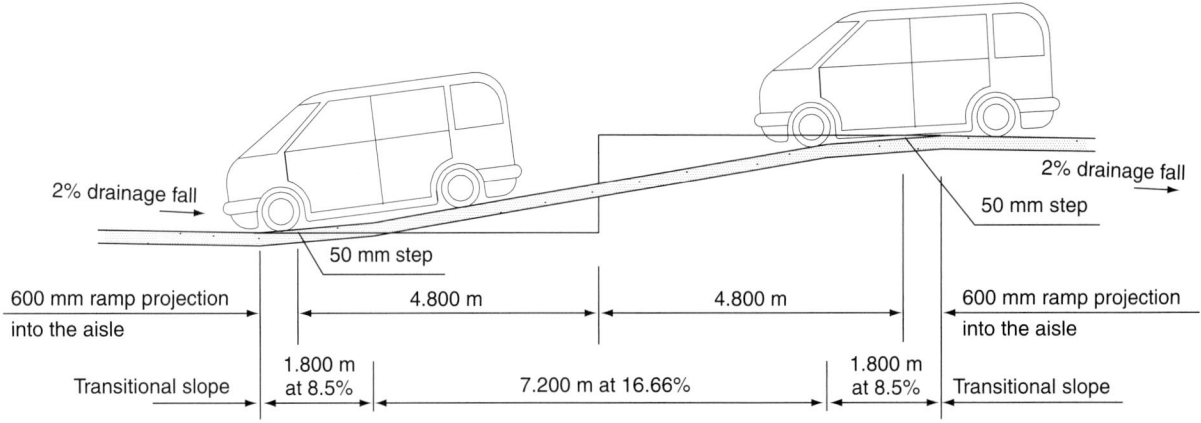

(b) Ramp based upon a 600 mm maximum aisle projection
(Ramp projections do not occur when storey heights are less than 2.720 m)

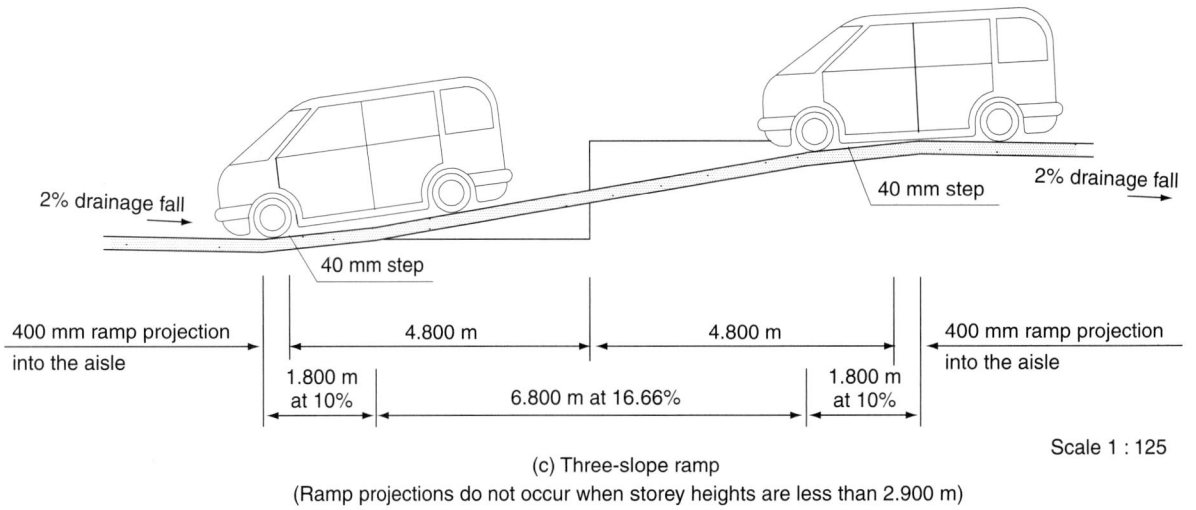

(c) Three-slope ramp
(Ramp projections do not occur when storey heights are less than 2.900 m)

Scale 1 : 125

Steps of more than 50 mm are awkward to 'feather' out without interfering with adjacent stall entries and they create obstructions to pedestrians pushing loaded shopping or luggage trolleys.

For good practice, ramp projections into an aisle should not exceed 600 mm in length and 50 mm in height.

3.5.5 Storey height ramps

Figure 3.9(a) shows the ramp shape for a straight 10% slope between storey heights of 3 m (FIR type) The recommended 'going' is 30 m and if used in a type 1 layout it will project 2.4 m into the aisle on each side.

Buildings of the FIR 1 type have been successfully constructed for many years without obvious complaint from motorists. It is, however, impractical to intrude 2.4 m into the side of an aisle and so another solution must be found if FIR-type layouts are not to be deemed unacceptable.

Figure 3.9(b) shows the ramp shape where the aisle projection is restricted to a maximum of 600 mm. The slope is 11.36% and has been used in existing layouts for many years with little complaint from motorists.

Figure 3.9(c) shows a 'conforming' ramp where the storey height has been halved and a 4-m-long landing has been introduced. It is, in effect, a pair of three-slope ramps end-connected at the landing level where the maximum half-rise does not exceed 1.5 m. The ramp intrusion into the aisle is eliminated. A three-slope ramp, however, will be a little less popular with motorists.

3.5.6 Side clearance

Minimum clearance dimensions of 3 mm should be provided on each side at the entry location into a cross-ramp or access-way.

3.5.7 Manoeuvring envelope

Figures 3.10, 3.11 and 3.12 show the manoeuvring envelopes (ME) for differing widths of ramps and cross-ways. The hatched area of the ME shows additional manoeuvring ability based on a 6-m radius turn and cannot be over-run without making a reverse manoeuvre. Between the boundaries motorists are able to decide on the route they will take and choose an appropriate rate of turn. The ME shows the degree of flexibility that an SDV motorist has when turning into a particular width of stall, ramp or access-way. Designers fitting, say, a car park into a basement, where the disposition of the structure may not be compatible with parking layout requirements, may have to compromise on preferred parking dimensions. It is useful to know the limits of the manoeuvres that can be made.

3.5.8 Stall access

Figure 3.10(a) shows the 'envelope' for access into a stall where the adjacent stalls are unoccupied. Drivers are able to overlap the upper stall and it shows that most vehicles can drive down the middle part of a traffic aisle, avoiding pedestrians and still gaining direct access into stalls located on either side.

Figure 3.10(b) shows a 6 m radius turn into a stall located between adjacent occupied stalls. Such a tight turn is acceptable for a single manoeuvre and with most vehicles being smaller than the SDV motorists can successfully keep within the limits of the ME without reaching their vehicle's limitations.

Figures 3.10(c) and 3.10(d) show that, without reducing parking standards, a reduction in the width of an aisle can be affected when wider stalls are used. It should be appreciated, however, that where clear span structures are involved and reducing stall widths is simply a matter of painting lines, future changes in the parking category could affect market values. When turning into a standard width stall, the aisle position is critical and an overshoot of more than, say, 200 mm will result in reversing manoeuvres being required. The wider the stall, the easier this manoeuvre becomes.

3.5.9 One-way flow ramp widths
3.5.9.1 Discussion

Figures 3.11(a) and 3.11(b) show the entry width into the ramps measured from the face of the aisle. Ramp projections can vary and therefore have not been shown. Designers should consider the effects that projections, steps and angled approaches will have on the layout. Circulation efficiency depends on the appearance that ramps present to the oncoming motorist. If constructed between solid sidewalls, even ramps of an adequate width are less inviting to enter, and drivers tend to be more cautious than they are when negotiating those where their lateral vision is unimpeded.

Figure 3.11(a) shows the manoeuvring envelope (ME) required to negotiate one lane of a pair of scissors-type ramps located between three parking stalls; the clear width between

Figure 3.9 Vehicle ramps with 3 m storey height

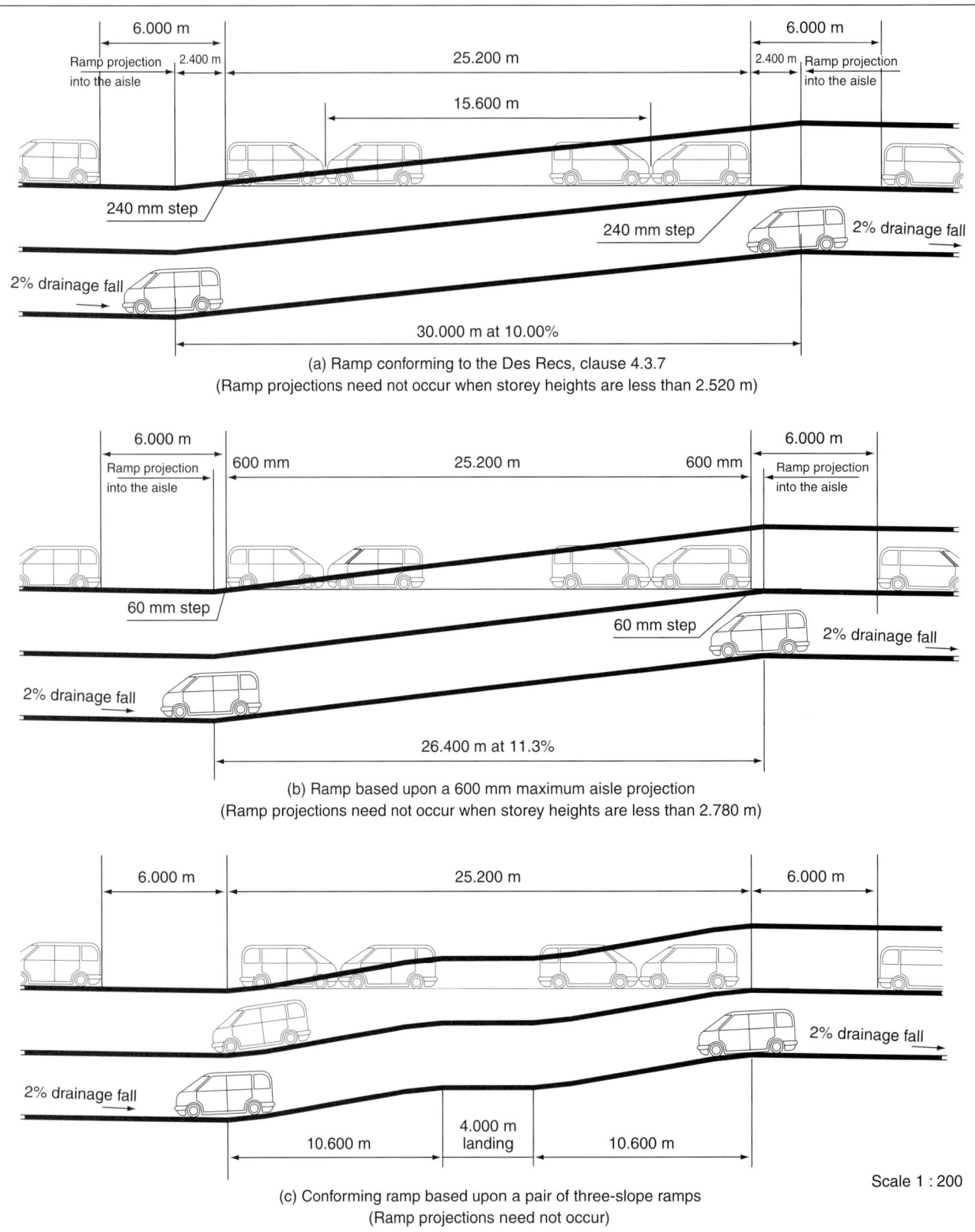

(a) Ramp conforming to the Des Recs, clause 4.3.7
(Ramp projections need not occur when storey heights are less than 2.520 m)

(b) Ramp based upon a 600 mm maximum aisle projection
(Ramp projections need not occur when storey heights are less than 2.780 m)

(c) Conforming ramp based upon a pair of three-slope ramps
(Ramp projections need not occur)

Scale 1 : 200

wall faces is 3.3 m. The lower lane will be a repeat of the upper lane and has been omitted for clarity but the potential conflict between adjacent vehicles at the exit with both turning in the same direction can easily be appreciated. A 6-m-radius turn is needed to access the cross-way that can be improved to 7.5 m on the exit. The turns are totally inadequate for a public 'user friendly' car park, although they might be acceptable in some private facilities where small cars are the norm.

Figure 3.10 Stall access manoeuvring envelopes based on 6-m-radius turns (read in conjunction with Section 3.5.8)

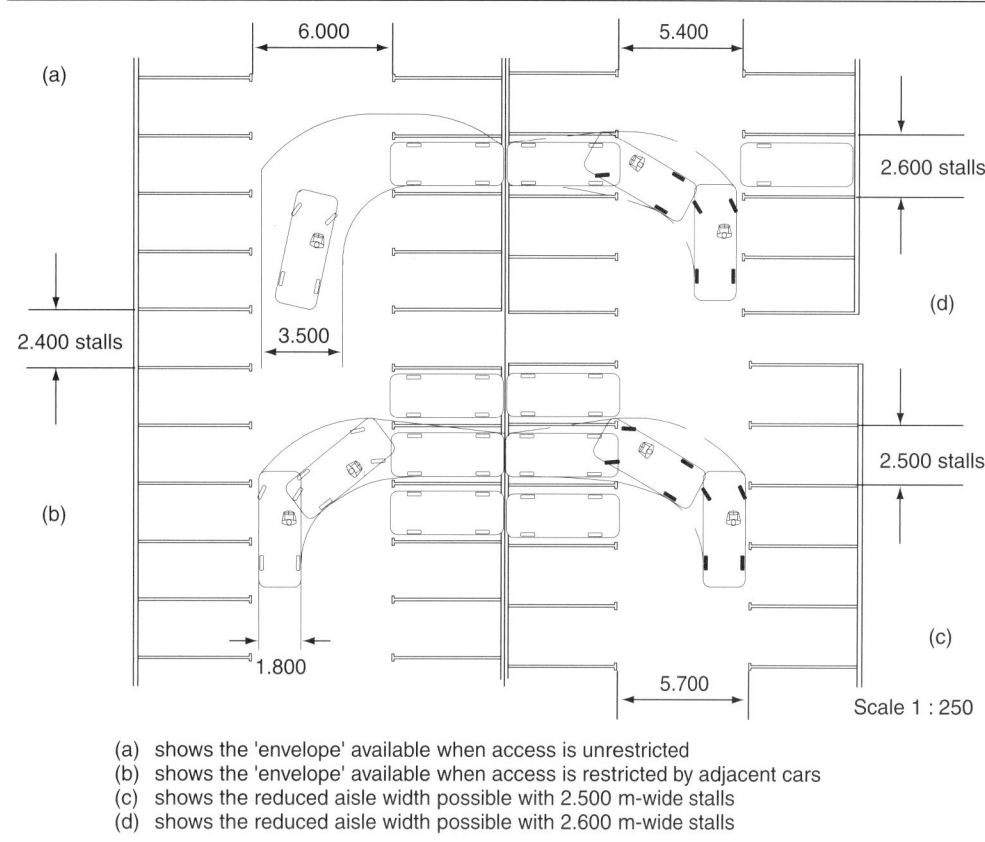

(a) shows the 'envelope' available when access is unrestricted
(b) shows the 'envelope' available when access is restricted by adjacent cars
(c) shows the reduced aisle width possible with 2.500 m-wide stalls
(d) shows the reduced aisle width possible with 2.600 m-wide stalls

Figure 3.11(b) shows the envelope required to negotiate a single ramp within a width of two parking stalls (4.8 m). The clear dimension between wall faces is 4.4 m and clearances of 300 mm have been incorporated on each side. A successful entry can be made with a 9-m-radius turn on both the entry and exit. The dynamic efficiency is high with vehicle speeds of 10 mph capable of being maintained throughout the turn. The upper shaded area shows the amount of over-run available if an entry/exit in a single manoeuvre is to be made possible. The lower shaded area indicates the aisle width available to drivers to achieve an entry or exit as a single manoeuvre.

As ramp widths increase, the aisle entry width also improves and at 6 m drivers can enter from virtually the complete width of an aisle. This will be highly popular, but the width of the Figure 3.11(b) ramp is an acceptable, economical, compromise, and is space-efficient. When ramp widths are excessively wide drivers can angle their vehicles as they approach the adjacent aisle. When located at the ends of the traffic aisle it is not important, but when located internally the manoeuvre reduces the driver's ability to observe approaching aisle traffic, especially when the passenger seat is occupied in a right-turning circulation system: for this reason, excessively wide ramps and cross-ways can have a detrimental effect on dynamic efficiency.

3.5.10 One-way flow ramps with 90° parking

Recommended clear widths between faces of structure

Upper limit	5.0 m
Minimum for good practice	4.2 m
Absolute minimum	3.6 m
Straight-ahead ramps	3.0 m

3.5.11 Ramp widths and angled parking

Where angled parking is adopted and the entry or exit does not involve a right-angled turn, some reduction may be accepted to the recommended figures, but the designer must

Figure 3.11 Ramp access manoeuvring envelopes based on 6-m-radius turns and one-way traffic flows (read in conjunction with Section 3.5.9)

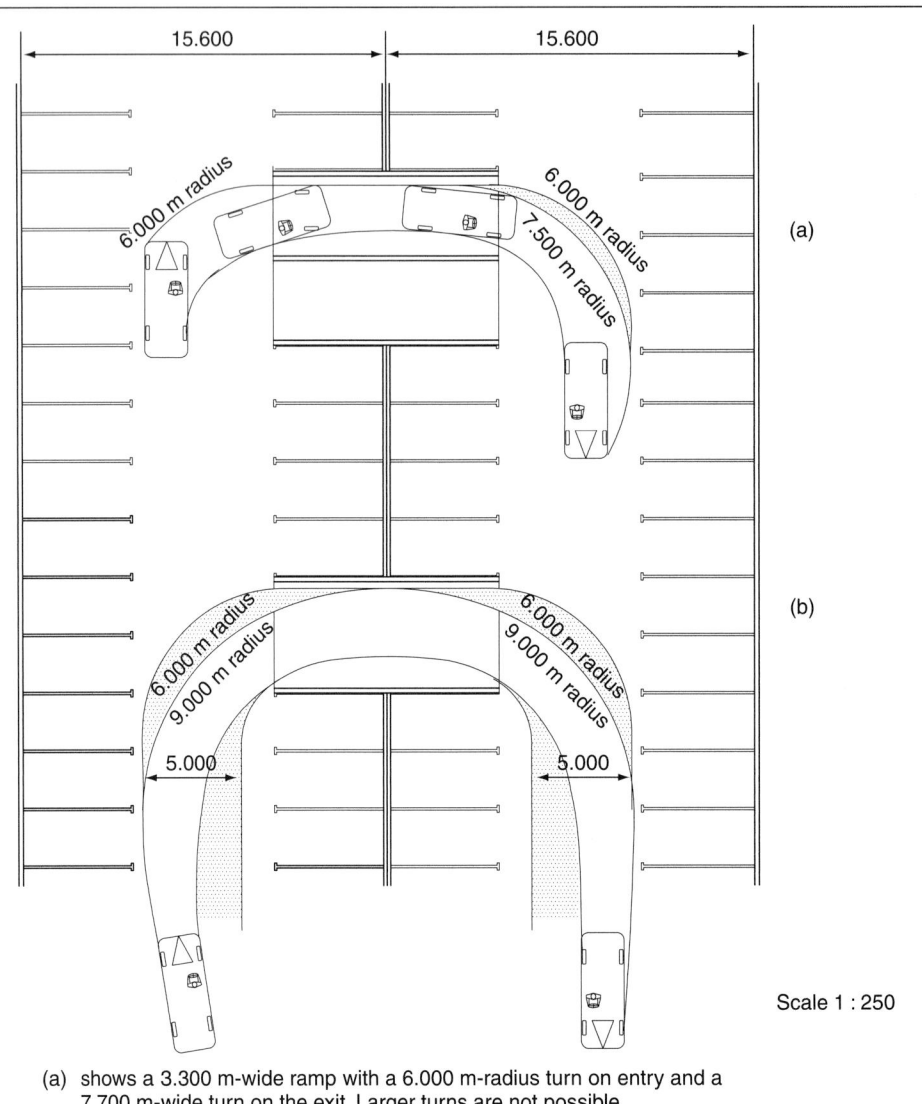

(a) shows a 3.300 m-wide ramp with a 6.000 m-radius turn on entry and a 7.700 m-wide turn on the exit. Larger turns are not possible
(b) shows a 4.400 m-wide ramp with 9.000 m-radius turns on entry and exit

Scale 1 : 250

consider the absolute necessity for such an action and the tolerance of the motorist who will use the facility. In any case, the minimum ramp width should never be less than 3 m.

3.5.12 Two-way flow ramps

Two-way flow ramps are generally designed three stalls wide (7.2 m), resulting in a clear width of about 6.8 m. They are used when it is desirable to achieve maximum static efficiency, mainly when the traffic flows in either direction are 'tidal' or not intensive. They are not recommended for large Cats 1 or 2 layouts but, when used in Cat. 4 layouts (staff parking type), they can be very effective with the traffic capable of using the full ramp width on entry (am) and on exit (pm). For this reason, painted lines are preferable to dividing kerbs, even for the tightest of turns. When two vehicles approach each other on the ramp, they are in danger of colliding at the exit and entrances. It is difficult for turns to be made without straying into the opposing lane and for this reason it is advantageous to angle the central median by about 300 mm each side of the dividing line to increase the width of the entry (see Figure 3.13).

3.5.13 Turning circle templates

Swept Paths for the recommended minimum (9 m radius) 90° and 180° SDV turns are shown in Figure 3.14 to a scale of 1:200. They can be photocopied to an appropriate scale and used for checking purposes.

Figure 3.12 Ramp access manoeuvring envelope (read in conjunction with Section 3.5.8)

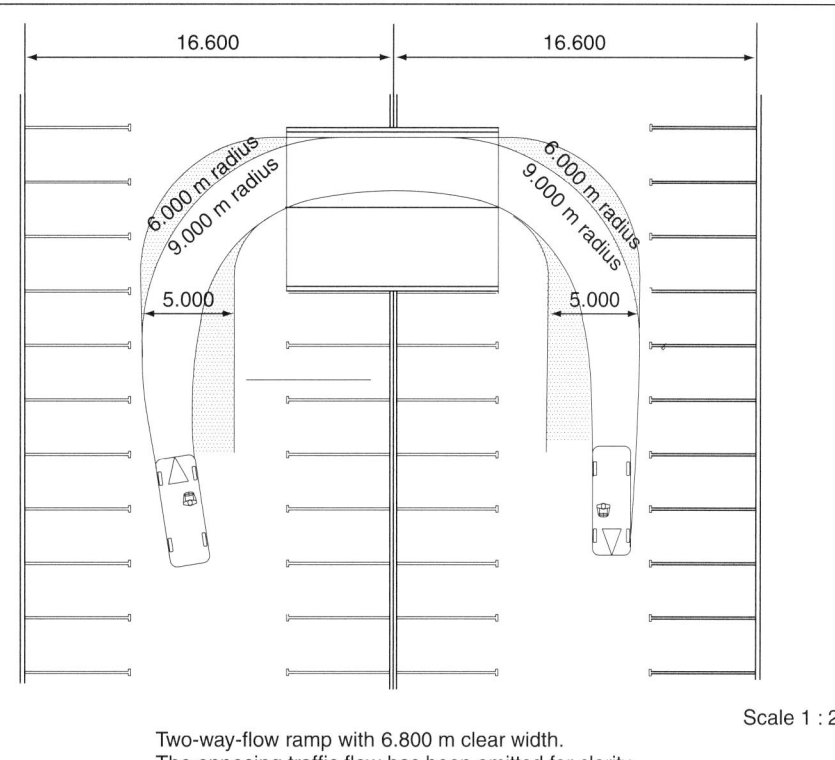

Two-way-flow ramp with 6.800 m clear width.
The opposing traffic flow has been omitted for clarity

Scale 1 : 250

3.5.14 Two-way flow recommended minimum clear ramp widths	Preferred	6.8 m
	Absolute	6.5 m

3.5.15 Scissor-type ramps

These ramps are generally designed to fit, in pairs, between three stall spaces in a one-way flow circulation system (see Figure 3.15). Having an overall width the same as a two-way flow ramp but with a central supporting structure, the resulting individual widths will be below the recommended minimum of 3.8 m. Nevertheless, they have been and are still being used in a number of buildings, mainly of the smaller 'private' type where the majority of vehicles can turn more tightly than the SDV (see SLD 3).

3.5.16 Side-by-side ramps

A fundamental problem occurs where vehicles on each ramp (one climbing and one descending) arrive at the same deck level side by side (similar in appearance to scissors ramps) and turn in the same direction, resulting in conflict between drivers. If intended for Cats 1 or 2 public car parks, there should be a minimum of two stall spaces located between them.

3.5.17 Circular ramps

These are less popular with the parking public than straight-ramp circulation systems. One of the reasons for this may well be that many have been constructed to a smaller diameter than the motoring public is willing to accept. There is a considerable difference between a vehicle's minimum turning circle and the minimum turning diameter that the average motorist will tolerate when committed to driving through one or more complete spirals. A 'wall of death' type sensation can occur in drivers when exiting through several floor levels in a tight turn. Eighteen metres, the recommended minimum diameter (given in Section 3.1.7) for turning on routes with straight aisles and cross-ramps, is not an appropriate dimension for circular ramps rotating through 360° or more. The constant turning and the inability for drivers to see any reasonable distance ahead renders greater diameters desirable if they are to be readily accepted by the motoring public. They are also difficult to supervise using CCTV cameras.

Figure 3.13 A two-way flow ramp

3.5.18 Recommended minimum diameters for full-circle ramps between limiting wall faces

One-way flow
For good practice   24 m
Absolute   20 m

Two-way flow
For good practice   31 m
Absolute   27 m

3.5.19 Recommended minimum widths for circular-ramp lanes, between wall faces

One-way flow
For good practice   4.4 m
Absolute   3.8 m

Two-way flow
For good practice   7.8 m
Absolute   7.2 m

3.5.20 Interlocking ramps

3.5.20.1 Stadium type

Where lengths of straight ramp connect semi-circular ramps, drivers can relax on the need for constant turning. Compared with continuous circular ramps they appear, visually, to be more spacious and hence less daunting. It has been noted that they remain acceptable to the driving public even when the diameters of the circular ends are somewhat below that recommended for full circular layouts. When the 'going' for a 360° rotation is better than about 56 m, another ramp, flowing in the opposite direction, can be inserted within the same plan form and can be a space-saving way of achieving access to a parking level located above commercial or retail premises (see ER 5 in Chapter 7).

3.5.20.2 Circular type

At a diameter of 24 m, a 10% gradient on the ramp centre-line produces a rise of about 6.2 m for each 360° rotation. Similarly, as for the stadium type ramp, another ramp can be introduced. When used in the 'interlocking' mode they take up less site area than a pair of circular ramps or a single two-way flow circular ramp.

3.6. Kerbs

Kerbs, historically, separate pedestrians on footpaths from vehicles. Pedestrian ramps should not exceed a slope of 5%, while vehicle ramps can slope up to 17%. Kerbs down the sides of vehicle ramps, therefore, are not intended for pedestrian use but to provide a

Design elements

Figure 3.14 9-m-diameter turning circles

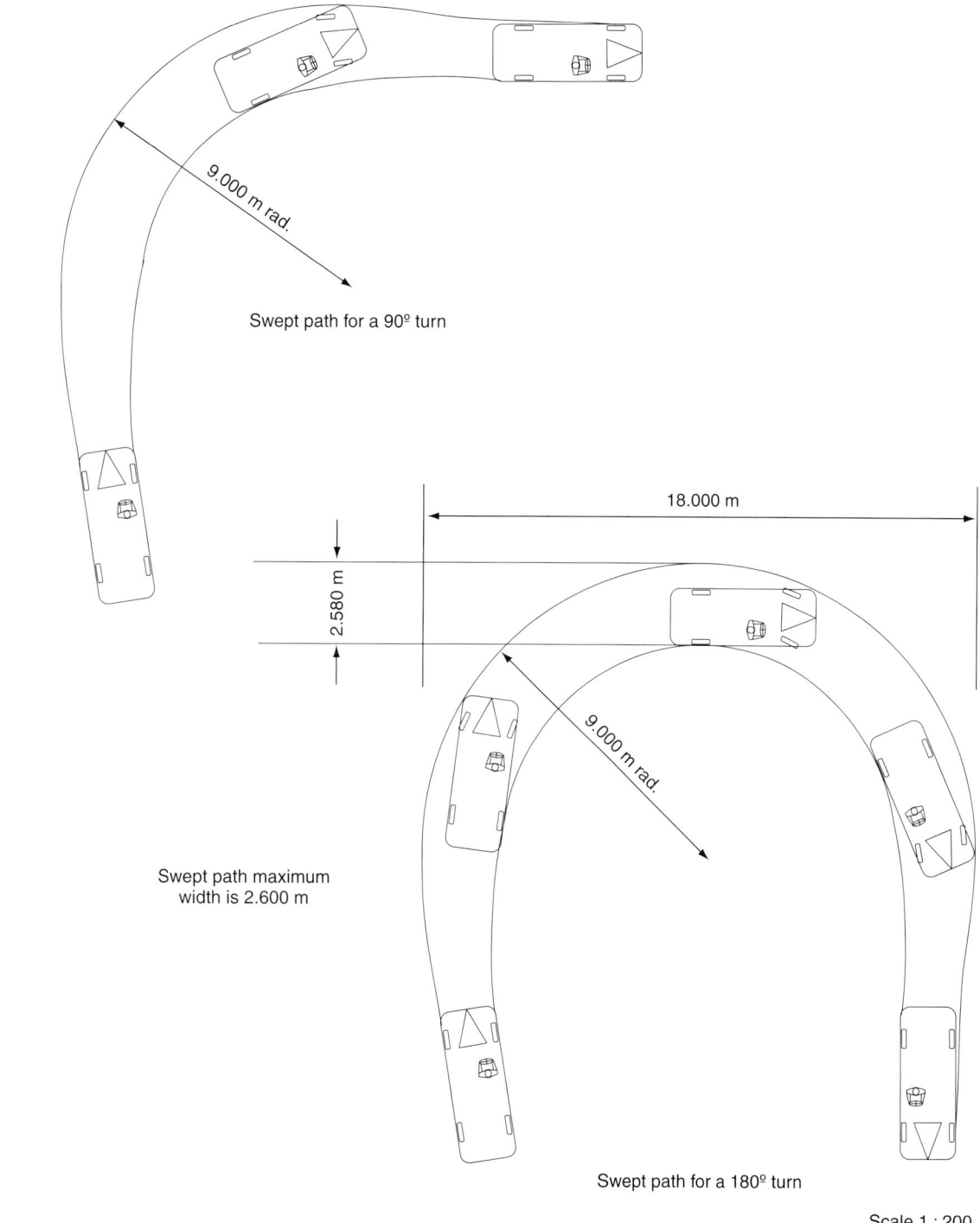

Swept path for a 90° turn

Swept path maximum width is 2.600 m

Swept path for a 180° turn

Scale 1 : 200

warning to motorists that the turning vehicle is getting close to a wall. The problem is that some pedestrians will see them as an invitation to access an adjacent parking level and use them accordingly. If kerbs are to be used, they should be of a width that does not encourage pedestrian use (not more than 600 mm wide).

If the 'preferred' minimum dimensions for car parking are observed, then the provision of kerbs may well be unnecessary and pedestrians will not be tempted to use them. It must be left to each designer to make a decision in this matter. Many conforming ramps have been constructed without kerbs and operate successfully.

Figure 3.15 Scissor type ramps

The provision of a kerb between lanes on a two-way flow aisle is an option that is required only when cars are travelling on the 'wrong' side of the ramp. When cars are driven on the correct side of a two-way ramp, the situation is little different from that of a circular car park, where lane-dividing kerbs are not used because access to the stalls is required from either side.

3.7. Super-elevation

Super-elevation of circular ramps is not a necessary feature in car parks where the maximum speed should not exceed 10 mph; in fact it is a positive deterrent to speeding.

3.8. Parking deck gradients

Where cars are parked sideways on a sloping deck, the maximum gradient should not exceed 5%. For pedestrians who also use the deck further restrictions may apply. Cars can be parked successfully on much steeper sideways slopes, but the effect of gravity on the opening and closing of their doors is a factor limiting the slope for use by the general public.

3.9. Headroom and storey heights

Headroom dimensions are often a compromise. They are frequently controlled by building height limitations for a particular site, coupled with the desire to incorporate as many parking levels as possible. Vertical circulation is a function of the storey height; the greater the dimension the longer and/or the steeper the ramp slopes become.

In order to contend with the height of modern vehicles, the minimum clear headroom throughout a parking building, measured under all light fittings, hanging signs and structure, should not be less than 2.1 m. It should be checked, especially at the bottom of ramps, where the wheelbase spans the change in slope and increases the apparent height of the vehicle.

It is important to predetermine the effect that down-standing light fittings, and signs, may have on headroom, particularly where flat soffites are involved, and also where ventilation ducts, fans and sprinklers occur in facilities constructed below ground level.

3.10. Height limitations

Height-limitation gantries should be located at the entrance to every car park, to prevent oversized vehicles from entering. They should be brightly painted and indicate clearly the maximum height of the vehicle allowed to enter the facility (see Figure 3.16).

Figure 3.16 A height-restriction barrier

Car Park Designers' Handbook
ISBN 978-0-7277-5814-9

ICE Publishing: All rights reserved
http://dx.doi.org/10.1680/cpdh.58149.027

Chapter 4
Dynamic considerations

4.1. Discussion

Dynamic capacity and efficiency is a measure of the rate at which traffic can pass any designated location, traffic aisle or even a complete circulation layout within a car park, and it enables potential bottlenecks to be identified. It can indicate the limitations of a particular layout and the need for change or improvement by the incorporation of by-pass routes or other special features.

4.1.1 Impact speeds

Calculations for the impact resistance of structural elements within a parking facility are based on a vehicle speed of 10 mph (BS 6180 (BSI, 2011a) and 6399), and should be assumed as the maximum permitted speed in any car park.

4.1.2 Effects of rain

In surface-only car parks and on the open top decks of multi-level facilities, wet and slippery conditions will occur from time to time. In such situations, drivers will be more cautious and dynamic capacities will be reduced. If a surface car park is operating at maximum dynamic efficiency when conditions are dry, it is likely that traffic congestion, queuing and delays will occur when it rains. For a structured car park, a covered roof may solve the problem. Designers should take these factors into consideration. On average, in the UK, rain occurs on one day in every three.

4.1.3 Exit and entry rates and internal movement

4.1.3.1 Discussion

Calculations for dynamic capacity should be checked against figures for the anticipated hourly vehicle movements but, in the absence of such information, it is generally accepted that 25% of a car park's static capacity should be able to enter or leave within a 15-minute period.

Short-term parking at busy supermarkets and major retail outlets builds up steadily from about 8 a.m, with peak activity occurring between 10 a.m. and 12 p.m. By that time, early customers are beginning to leave and others are slowing down to search (and sometimes wait) for spaces in a favoured location, for example, adjacent to a lift or shopping access lane. The effect that pedestrians, some with prams and others with shopping trolleys, have on traffic movements has not been thoroughly investigated, but it has been observed that in facilities where this occurs, dynamic capacity can be affected, adversely, to varying degrees, depending on the parking standards encountered.

Figures on dynamic capacity, proposed in RRL report LR 221 (Ellson, 1969), are based on the assumption that an aisle fills in a logical manner, starting at the beginning and finishing at the end, with no cars leaving as it fills, no cars arriving as it empties and no pedestrians on the traffic aisles. This report formed part of the first edition of the 1976 'Des recs' but was omitted from subsequent editions.

Observations made in a number of car parks indicate that the RRL figures for dynamic capacity can be applied, within reason, to Cat. 3 and 4 car parks. For the others, however, there is evidence that a reduction is justified.

Where two-way flow aisles are used, an additional problem can occur when vehicles, entering and leaving stalls, cross the oncoming lane and interfere with traffic travelling in the opposite direction. When occurring in facilities with mainly tidal flow, dynamic capacity will be similar to the figures provided for one-way traffic flows, but for short-stay, intensive use facilities, it is affected adversely.

In large car parks, the build-up of traffic from successive upper parking levels can be greater than the dynamic capacity of the lower levels of an exit ramp. It must also be appreciated that if the exit control or external road system is unable to cope with the flow rates, traffic congestion will occur within the car park, regardless of any other factor.

4.1.4 Dynamic capacities for different stall widths and categories

4.1.4.1 Discussion

The figures given are averaged out from the result of observations made, over 15-minute periods at peak times, in a number of car parks where the exiting traffic rate was not reduced by external conditions. The variations between them were such that precision is not a factor. They are, however, considered to be a conservative but realistic assessment. The variations between different car parks were attributed mainly to the dynamic efficiency of the individual layouts.

Staff parking tended to be more rapid than the other types and it proved impossible to obtain meaningful '15-minute' figures from Cat. 3 long-stay car parks.

The difference between the inflow and outflow figures can be explained by the fact that most motorists can turn and drive straight into a stall, but when exiting, a more hesitant reversing manoeuvre is involved. A small proportion of drivers, however, reverse into the stalls in order to drive straight out.

Notional figures for all parking categories with 6-m-wide aisles

	Inflow	Outflow
2.3 m	820	710
2.4 m	860	750
2.5 m	910	800.

4.1.5 Stopping distance

An extrapolation of the stopping distance, given in the Highway Code, for cars travelling at 10 mph on dry roads, is about 6 m. Designers must be aware, however, that wet, exposed decks can double the stopping distance..

4.1.6 Speed limits

There are no national regulations governing vehicle speeds within a car park, but it has become established practice to adopt the same speed that the structure must be designed to withstand (10 mph). Some authorities, however, are proposing speed limits of 5 mph. If adopted nationally, the traffic-flow numbers used in dynamic design for aisles, and ramps should be reduced accordingly: congestion will occur at a much lower figure than the recommended maximum search path of 500 stalls ('Des recs' 4.4.7) and should be adjusted downwards. Dynamic design has, historically, been based on speeds of about 10 mph and the figures for inflow and outflow given in Section 4.1.4.1 are based on observations made in car parks where traffic has not been restricted to 5 mph. It is unrealistic to impose lower vehicle speeds and yet still expect the original circulation efficiency to be achieved..

4.1.7 Dynamic capacities of ramps and access-ways

4.1.7.1 Discussion

At a maximum speed of 10 mph (16 km/h), keeping the correct stopping distance between vehicles and assuming that all vehicles are the same length as the SDV, it can be calculated that the unobstructed vehicle rate should not exceed about 1450 cars per hour. Many cars, however, are shorter than the SDV and a more realistic figure can be based on an average vehicle length of 4.3 m, in which case the allowable figure rises to about 1500 per hour. It would be imprudent to design a car park using flow figures that are greater than those developed from conforming data. In the event of a traffic accident, designers should be able to demonstrate that they did not use figures that enabled motorists to drive in excess of the maximum 'recommended' speed.

The rate of entry into a cross-ramp or access-way depends on its width and the appearance it presents to the motorist. Above a clear dimension of 4.2 m, the 1500 vph (vehicles per hour) rate does not appear to be affected, but, as the entry width reduces, drivers become more cautious and tend to slow down as they turn. The situation is also exacerbated by sidewalls, or other visual obstructions (see Figure 3.15).

4.1.8 Dynamic capacities of cross-ramps and access-ways, per hour

Observations of the effect that ramp widths have on motorists indicates that for driving on ramps free of lateral obstructions to vision, the following dynamic capacities can be recommended.

>4.2 m wide   1500
3.6 m wide   1200

Straight-entry ramps
3.0 m wide   1500

Under 3.0 m wide, no information is available, but it is reasonable to assume that a progressive reduction will occur down to about 2.7 m when most drivers will refuse to enter. In some large-capacity, short-stay facilities, the exit rate could exceed the traffic capacity of the external road system. It is often prudent to check this condition before reaching a final decision on the design.

4.1.9 Dynamic capacities of parking decks
4.1.9.1 Calculations

From 4.1.4.1, the notional dynamic capacity of a 6-m-wide aisle can be seen to be 860 and 750 cars per hour respectively, for inflowing and outflowing traffic. The dynamic capacity of the ramps is constant at 1500 cars per hour.

The calculations are based on the premise that, without parking traffic, it would be possible for 1500 vehicles per hour to progress on the traffic aisles regardless of the length of travel. Cars entering and leaving stalls, however, will reduce this speed and, hence, dynamic capacity, dependent on the number of stalls involved. A very few stalls will have little effect, but a large number could slow down the flow rate to a figure approaching that given for the notional dynamic capacity.

Regardless of aisle length and capacity, the one-hour flow rate cannot be less than that given in 4.1.4.1 or greater than that given in 4.1.8. It can be expressed by the formula

Actual dynamic capacity (ADC) = $1500 - [a \times b \times c \times 1/d]$

where a is 1500 minus the notional dynamic capacity, b is the number of flanking stalls divided by a, c is the number of aisles, d is the stall turn-over rate.

Example
Calculations for a Figure 4.1, Cat. 1-type layout with 2.4-m-wide stalls (four storeys, ten aisles and a 1.5-hour stall turnover.

Figure 4.1 A Cat. 1, SLD2 car park

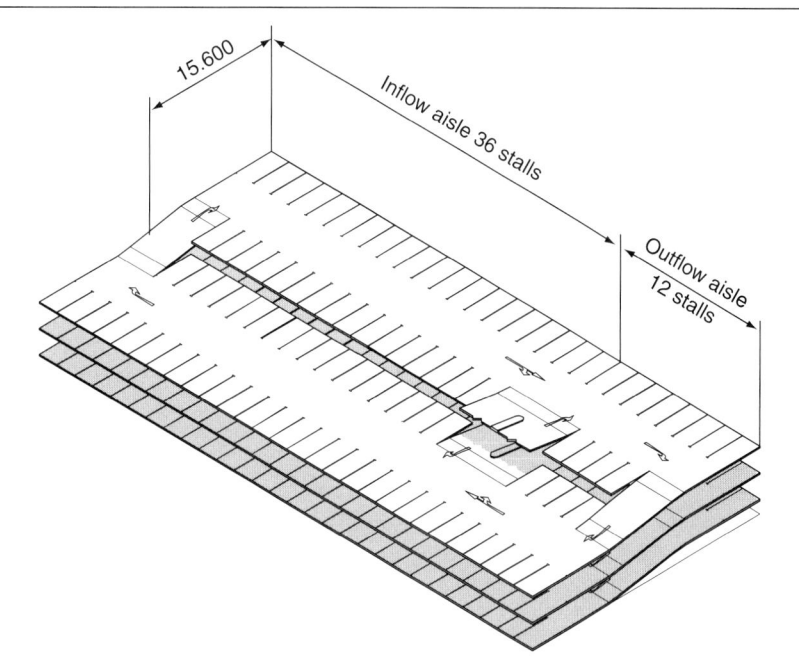

Inflow
Aisle 1, Level 1 (36 stalls)

$$\text{ADC} = 1500 - [640 \times 36/640 \times 1 \times 1/1.5] = 1476 \text{ vph}.$$

At the end of the inflow route, having passed through 10 aisles, ADC will be reduced by 240 to 1260 vph.

For outflow routes, the calculations need to be made from the top deck down.

Outflow
Aisle 10, Level 5 (12 stalls)

$$\text{ADC} = 1500 \times [750 \times 12/750 \times 1 \times 1/1.5] = 1490 \text{ vph}$$

At the end of the exit route, having passed through ten aisles the ADC will have reduced to 1400 vph.

Although this is a small example, it can be appreciated that when much larger facilities are involved and aisles need to be driven through more than once, the calculation can be used to determine where congestion is likely to occur and where an alternative route can be used to advantage.

4.1.10 Dynamic efficiency

Angled parking is more efficient, dynamically, than right-angled parking when both incorporate stalls having similar dimensions. As the angle reduces, so the stalls become easier to enter and leave, and it becomes increasingly difficult for cars to turn against the traffic flow. This improvement results, generally, in a reduction in static efficiency, a narrowing of the aisles wherein traffic and pedestrians mingle, and a reduction in the distance available for turning between the outside faces of adjacent aisles.

There is a case for parking at angles down to 70° on the internal bins in a multi-bin configuration (Figure 3.3), where the static and dynamic efficiency is slightly superior to that required for 90° parking. It must be remembered, however, that as the parking angle reduces the stall pitch increases, and this could be detrimental to the static capacity of a facility with a fixed overall length.

Car Park Designers' Handbook
ISBN 978-0-7277-5814-9

ICE Publishing: All rights reserved
http://dx.doi.org/10.1680/cpdh.58149.031

Chapter 5
Static considerations

5.1. Static efficiency: discussion

The static efficiency of a car park is a function of its static capacity and the area of the parking decks. It is used as a means of comparison between parking facilities and is couched in general terms such as 'Good', 'Average' or 'Poor'. Large-capacity decks, where the ratio of parking spaces to total floor area is high, will produce better figures than small-capacity facilities, where higher percentages of the parking decks are given over to ramps and access-ways. However, the terms are relative to the most efficient layout that can be achieved for any particular deck capacity (see Figure 3.3 and Section 3.4).

For example, a 300-space-per-deck layout requiring 28 m^2 for each car space can be described as 'Poor', since it is possible to achieve a figure of 20 m^2 with an efficient layout. Conversely, a 30-space-per-deck layout requiring 28 m^2 for each car space can be described as 'Good', since it is about as efficient as it is possible for it to get.

5.1.1 Relative efficiencies

A long two-bin layout has a greater static efficiency than a shorter three- or four-bin layout of a similar floor area, due to the reduction in the number of access-ways required for access between adjacent stalls, each of which takes up the space of four stalls at the ends. It can be a useful consideration when deciding on the layout for a new car park.

Floor area requirements for different angles of parking, as a ratio compared with that for 90° parking, are provided in Tables 5.1, 5.2 and 5.3. They are based on stall widths of 2.4 m and one-way traffic flows (see also Figure 3.3).

Column A shows the pitch of the stalls in metres.
Column B shows the bin width in metres.
Column C shows the percentage area variation.

Table 5.1 Single bins

Angle	A	B	C
90°	2.400	15.600	100.0
80°	2.437	15.530	101.0
70°	2.554	15.362	105.0
60°	2.771	14.914	111.0
50°	3.132	14.240	119.0
45°	3.394	13.782	125.0

Table 5.2 External bins

Angle	A	B	C
90°	2.400	15.600	100.0
80°	2.437	15.328	100.0
70°	2.554	14.952	102.0
60°	2.771	14.314	106.0
50°	3.132	13.469	112.0
45°	3.394	12.939	117.0

Table 5.3 Internal bins

Angle	A	B	C
90°	2.400	15.600	100.0
80°	2.437	15.120	100.0
70°	2.554	14.541	99.0
60°	2.771	13.714	101.0
50°	3.132	12.697	106.0
45°	3.394	12.085	110.0

Two-way flow layouts can only be used sensibly with 90° parking.

Compared with the area for one-way flow 90° parking, the (Column C) figure for two-way flow is 106.4%.

5.1.2 Area per car space

As a guide, the floor areas per car space that can be termed 'Good', for five different deck capacities with 90° parking, including a reasonable allowance for stairs, lifts and so on, are shown in the following table.

300 stalls per deck	20 m²
200 stalls per deck	21 m²
100 stalls per deck	22 m²
60 stalls per deck	24 m²
30 stalls per deck	28 m²

For angled parking, these can be multiplied by the Column C figures shown in Tables 5.1, 5.2 and 5.3.

Significant variations can occur, especially where awkward-shaped sites are involved and the figures should be used as no more than a guide to the areas per car space that can be achieved under suitable conditions.

5.1.3 Recommended capacities

For good practice, the maximum static capacities of various car parks having circulation efficiencies better than 60% should be of the following numerical order.

5.1.3.1 Parking categories 1 and 2

Combined one-way flow routes
SLD 1 type   400 spaces

Separated one-way flow routes
SLD 2 type   600 spaces

Two-way flow
SLD 3 type   600 spaces

Extended one-way inflow route with a separated rapid outflow route
VCM 1 type   800 spaces

Extended and rapid inflow route with a separated rapid outflow route
VCM 1 type   1000 spaces

Flat decks with half external rapid outflow routes
HER types   1250 spaces

Flat decks with fully external ramps
ER type   1500 spaces

5.1.3.2 Parking category 3

Long-stay, non-tidal 'main terminal'-type car parks, where the traffic flow is not anticipated to reach the dynamic capacity of the traffic flow routes, need not be restricted in static capacity. However, rapid flow routes that enable drivers to reach any part of the layout within five minutes should be incorporated.

5.1.3.3 Parking category 4

Tidal flow layouts have at least two peak vehicle movements per day. If it can be anticipated that the peak rates will not exceed 25% of the car park's capacity over a 15-minute period, the cats 1 and 2 figures can be used.

Where layouts with circulation efficiencies less than 60% are considered, they can be compared with the 60% recommendations and their capacities reduced proportionally.

Car Park Designers' Handbook
ISBN 978-0-7277-5814-9

ICE Publishing: All rights reserved
http://dx.doi.org/10.1680/cpdh.58149.035

Chapter 6
Circulation design

6.1. Discussion

Parking layouts are either inhibited or uninhibited depending on the influence that other disciplines have on them. An inhibited layout is where parking has to fit between a pre-determined disposition of vertical services and structure, such as the basement car park of an office building. In such situations, designers have little opportunity to develop a fully effective layout and must do the best that they can. An 'uninhibited' layout is one where the designer has a 'clean slate' on which to work; where the only limiting factors are those of static capacity and the dimensional restrictions imposed by the site. The layouts described in Chapter 7 are of this type.

Surface car parks requiring a high static capacity will generally follow the tenets contained in Chapter 3, modified as necessary to follow the contours of the site and other geographical features. Such layouts are, invariably, one-off designs and it is impractical to provide examples other than as general recommendations.

6.2. How many levels?

There is no technical limit to the number of suspended levels a structured car park can have, but it is generally accepted that six is a reasonable maximum for it to be freely accepted by the motoring public and operators. Factors such as the intensity of demand, the availability of a suitable site and, of course, the requirements of the relevant planning authority will also influence this decision (see Figure 6.1).

Figure 6.1 Ten-storey car park above a restaurant in Leeds

In the USA, car parks have been constructed with more than 12 parking levels although, on average, the number of floors in most buildings is not dissimilar to UK practice; there are no fixed rules in this matter and much depends on public demand, the skill of the designer and the tolerance of the motorist.

Current commercial demands and land availability in the UK are tending towards taller facilities on smaller plot sizes.

6.3. Roof considerations

Most car parks in the UK are open to the elements. Roofing over the top parking deck occurs in only a relatively few cases. There are arguments for and against protecting the top deck with a lightweight roof.

- A waterproof membrane on an exposed top deck will require substantial renewal at least twice during a projected life of 60 years. It will also require expenditure on maintenance from time to time. Extreme inclement weather will tend to reduce the use of the exposed top deck.
- A lightweight roof over a top parking deck will cost about three times more than that for a single application of a waterproof membrane and, without attracting significant maintenance costs, should last the life of the building. It will also enable normal parking activities to be carried out in extreme weather conditions.
- An open-deck building can be constructed, initially, at a lower cost than a roofed-over car park and if the difference was invested over a projected life of some 60 years, the overall costs would be similar.
- Where the overall height is limited and the static capacity needs to be as high as possible, it makes sense to utilise all of the available building height for parking and construct a roof-top parking deck. If inclement winter weather eliminates roof parking for a time, it will still have the static capacity of a roofed-over car park in that particular location.
- Where it is reasonable to construct a lightweight roof without reducing static capacity, the advantages are that the building remains dry at all times, the top deck parking is not weather sensitive, protection from the summer sun is provided and long-term maintenance costs are reduced, if not eliminated altogether. Also, the building is less prone to structural deterioration and will have an enhanced market value.

6.4. Circulation efficiency
6.4.1 Discussion

In some car parks, the circulation design enables most or all of the stalls to be searched with just one circuit of the aisles and access ways. In other car parks, however, aisles must be driven through more than once to achieve a similar result. It is a factor worthy of consideration and affects dynamic efficiency, as well as parking times, especially in large capacity, multi-bin layouts. It is a complex problem to solve precisely since much depends on whether the car park is 'empty and filling', 'full and searching' or 'full and emptying'. The use of 'variable message signs' also affects circulation efficiency, since they enable drivers to by-pass aisles that are full and drive more effectively to an available stall.

The object is not one of precise assessment but rather one of establishing the relative circulation efficiency for one layout and comparing it with that for another. Provided that both are assessed in the same way comparisons can be made without undue complexity.

6.4.1 Shortest travel distance

The shortest travel distance possible to pass stalls located on each side of a traffic aisle is $2.4/2 = 1.2$ m per stall and can be equated to a circulation efficiency of 100%. It can only be achieved in a single-bin facility where motorists enter at one end and exit at the other end. Where cross-ramps and access-ways are used to complete the circulation in multi-bin layouts the circulation efficiency will be reduced and will vary according to the chosen layout design.

6.4.2 Examples of circulation efficiency

Example 1
Reference to the SLD 1 layout on pages 43 and 44 shows that all 96 of the stalls on each deck, plus getting to the next upper storey, can be achieved with a single circuit consisting

of four right-angled turns. Measuring along the centre of the aisles and access-ways the travel distance is

$$52 \times 2.4 \text{ m} = 132.8 \text{ m}$$
$$2 \times 15.6 \text{ m} = 31.2 \text{ m}$$
$$\text{Total distance} = 164 \text{ m}$$

Divided by the number of stalls the travel distance per stall is 1.7 m producing a circulation efficiency of $1.2/1.7 = 70\%$.

Example 2
Reference to the FIR 1 layout on page 85 shows that three circulation options are available.

Option 1
A single circuit, with four right-angled turns and climbing to the upper deck level, passes 58 stalls out of the 108 on each deck. It produces a circulation efficiency of 53% for the stalls passed and is only really suitable for getting quickly to the upper parking levels. Even so, that route is not very rapid.

Option 2
Including the stalls on the central aisle will increase the number to 82, but entails driving twice through one of the aisles and making eight right-angled turns. The efficiency of this route is 43%.

Option 3
To cover all of the stalls on each deck before driving up to the next level involves passing twice through one aisle, three times through another, and making 12 right-angled turns. The efficiency reduces to 33%

It can be seen that motorists can spend up to twice the time searching for an available stall in the FIR 1-type layout than in a VCM 1 layout.

Poor circulation efficiency is a major factor in creating traffic congestion and one of the main reasons why some car parks are less popular than others.

6.5. Parking times
6.5.1 Discussion

Five minutes is about the maximum time that an average driver is willing to spend searching for a stall in which to park, beyond which dissatisfaction and frustration with the building begins to develop. It can be a factor in deciding a motorist's future parking destination. There are reports of car lights left on, boot lids left open, even drivers who have left their car doors open with the engine still running in their panic not to miss an appointment or catch a plane or a train. Poor circulation efficiency and the frustration it causes can be a contributory factor in creating such situations.

The decision whether or not to alter the layout or incorporate rapid inflow routes can be influenced by an assessment of the time it takes to search all of the stalls. In the absence of more accurate information it is normal practice to assume that the peak flow rate in either direction will be 25% of the static capacity in any 15-minute period; this rate can be applied, within reason, to most single and multi-level parking layouts.

The notional inflow capacity of an aisle with 2.4-m-wide stalls is 860 vph. This produces an average speed just under 10 km/h or 3 m per second. However, the time spent on access-ways and ramps connecting adjacent aisles slows the average stall searching time to about 2.4 m per second. It is not a precise figure and so does not justify recalculation for minor differences created by varying stall widths. It should be used simply to establish the approximate times it takes to reach various parts of a car park and to compare the relative efficiencies of different circulation layouts.

Application of the five-minute recommendation limits simple 'follow-my-leader' layouts, with circulation efficiencies better than 60%, to a maximum of about 600 spaces and correspondingly smaller capacities for less efficient layouts. If, however, a speed limit of 5 mph is adopted for any particular building, the vph figure should be reduced to about 360 spaces.

In many large capacity layouts, mainly of the SLD and VCM series, rapid exit routes form part of the basic design, but rapid inflow routes can also be incorporated. Passing as few as 24 stalls for each storey height, the introduction of such routes enables drivers to by-pass full or congested lower decks and reach the emptier upper parking levels without exceeding the preferred maximum search time.

Car Park Designers' Handbook
ISBN 978-0-7277-5814-9

ICE Publishing: All rights reserved
http://dx.doi.org/10.1680/cpdh.58149.039

Chapter 7
Circulation layouts

7.1. Discussion

Of the more than 7000 structured car parks believed to have been constructed in the UK alone, it can readily be appreciated that no single person can have knowledge of every circulation layout variation that has been proposed and built. Practical considerations, personal experience, and the constant pressures for financial economy render it reasonable to assume that the examples shown, all of which have been featured or built during the past 50 years, provide the basis for most of the self-parking buildings that exist at the time of writing. The design of a satisfactory circulation layout is one of the most important factors governing user appreciation and yet many designers are unaware of the large variety of options from which they may choose and their suitability for the intended purpose.

The following examples are all practical layouts and form the basis upon which most self-parking facilities have been designed. Some are more popular than others and some are significantly defective in circulation design, static and dynamic efficiency. If designers are to gain confidence in developing solutions to solve particular problems, then it is desirable that they should know the strengths and weaknesses of individual layouts in order to make an informed choice.

7.2. Dimensions used

There are few precise dimensions that must be adopted for the design of parking structures. Dimensions for the individual elements can vary and are also affected by the parking angle (that varies according to the bin width) in one direction and the stall pitch (that varies the overall length) in the other direction. The main concern is that motorists and clients are content.

It is overly laborious and unnecessary to keep mentioning all of the variations that can occur in practice and so dimensions for the featured layouts will be based on those recommended for 90° parking with stall dimensions of 2.4 m × 4.8 m, aisle widths of 6 m (one-way flow) and 7 m (two-way flow), with a storey height of 3 m.

In the layouts shown in the following pages, the overall aisle lengths are sometimes shown less than those given for the width; nevertheless, the length of the aisle will determine the 'length' of a layout and the dimension over the bins will determine its 'width'.

7.3. User-friendly features
7.3.1 Discussion

There are many existing car parks where, in retrospect, it can be seen that the layout would have been much better if only the designer had recognised that a problem existed. In such cases, if improvements had been incorporated at the design stage, they need not have cost more to implement, or reduced the static capacity; they could even have enhanced the market value by being more 'user-friendly' to the parking public. It is, also, a relatively simple matter to spoil a potentially acceptable circulation layout by over-complication, or by the introduction of unnecessary and unfriendly features.

7.3.2 Simplicity

The basic tenet of all circulation design is to 'keep it simple'. What, at first, might look like a clever idea to a designer could well end up as a motorist's nightmare. In a structured car park the layout should endeavour to replicate the openness of a surface car park. To this end, it is desirable to eliminate, as far as possible, vertical structure that interferes, both visually and physically, with the free movement of vehicles and pedestrians. Turning directly from one lock to the other is not a popular manoeuvre. If possible all turns should be in the same direction and not more than 90° at a time. When located under

39

other types of building, it is not always possible to create the most desirable layout. Attempts should be made to minimise the visual impact of large vertical elements and locate them away from the circulation routes, if at all possible.

7.3.3 Cross-overs

Cross-over conditions should be avoided. When on a traffic aisle and searching for the first available space, it is disconcerting and potentially dangerous to find a car suddenly appearing at right angles from behind a parked vehicle. The driver of this car may also be concentrating on finding a space in which to park, or intent only on leaving the facility as quickly as possible. A 'user-friendly' circulation layout should not hold surprises for drivers, who should be able to observe the movements of other vehicles well before there is a need to take avoiding action.

7.3.4 Circulation direction

The direction of circulation has little effect on dynamic efficiency in one-way flow systems. Provided that the route is of an adequate width it matters little in which direction the traffic is made to flow. It has been said that left-turning circuits are not as popular in one-way-flow systems as turning to the right. However, when vehicles are travelling down the aisle drivers are biased to the right thereby providing a slightly better view of openings on the left.

When drivers enter and exit a facility side by side, turning left to avoid a cross-over with exiting traffic directs them into a right-turning circuit. This in turn means that any side-by-side internally located ramp must be entered on the right. Provided that the entry is of any adequate width and properly signed it has proved to be no problem to driving efficiency.

Right-turning onto an exit-pay station from a right-hand-drive vehicle ensures that a ticket can be inserted more easily into the acceptor machine than when the driver is turning to the left, and is thus on the outside of the turn. None of these points are important enough to dictate the direction of flow by themselves, but it is useful to appreciate that they occur, when considering the flow direction.

7.3.5 Dead ends (culs de sac)

When viewing down a 'dead end' aisle, it is difficult to see the parking situation more than three or four stalls away. For good practice, and if unnecessary manoeuvring is to be avoided, this should be the limiting factor.

7.4. Angled and right-angled parking, a comparison

Members of the public and some clients ask why angled parking is not used more frequently in the UK. They point out that in the USA it is a popular parking format and makes it easier to park. However, in the UK, layouts with 90° parking occur more often than any of the other angles. There are two main reasons why this is the case.

First, in the UK, most drivers have grown up with flat-deck SLD-type parking, and with stalls at the most economical 90° angle. Most UK experience of sloping parking decks stems from a relatively rare incidence of two-way flow, CSD- and SD1-type layouts with two-way flows and 90° parking, where 'Up' is the inflow route and 'Down' is outflow.

The second reason is that sloping deck layouts with one-way flow require exiting vehicles on the inflow route to continue to climb in order to get to an exit route. If drivers were to park at 90° they would be tempted to exit 'downhill' against the designed traffic-flow. Parking at a lower angle, usually 75–80°, prevents that occurring. In the USA, a large proportion of car parks are of the SD and FSD types with one-way flows and they, invariably, employ angled parking. For two-way-flow layouts, however, parking is always at 90°.

Figure 7.1 shows a basic split-level, 90° parking layout, compared with three other parking angles where the aisle widths have been developed from the need for drivers to exit onto the aisle in one movement (see Section 3.4). The differing area requirements can be seen and they show that 90° parking layouts are not only the most statically efficient, but they also have an aisle width that provides the greatest separation distance between vehicles and pedestrians on the aisles.

Comparing 90° to lesser-angle layouts, it can be seen that as the parking angle reduces the building length increases and the aisle widths narrow. At a parking angle of 45°, a

Figure 7.1 Comparison between different parking angles

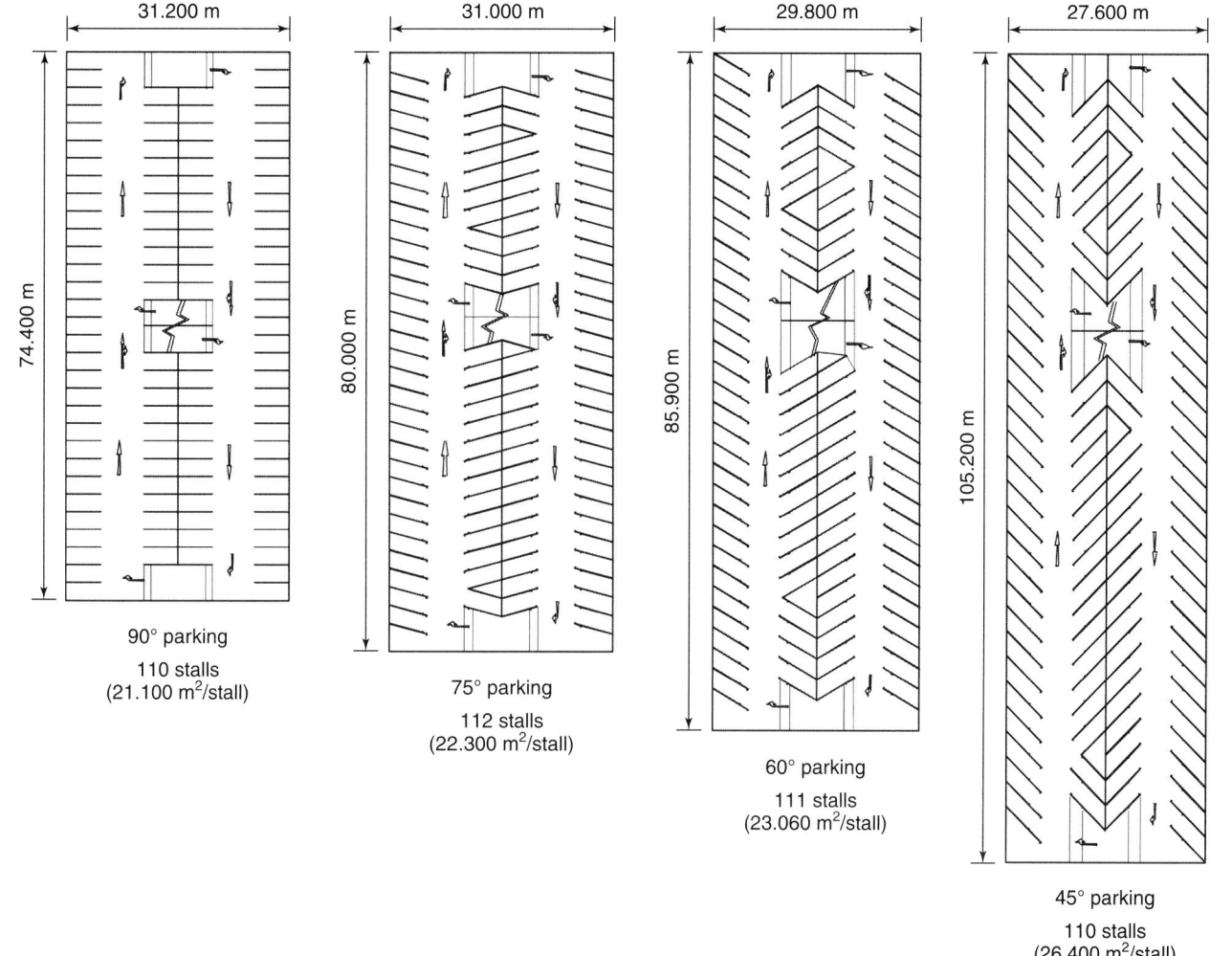

110-stalls-per-deck building will need to be 105.2 m in length (41% longer), and even with reduced aisle widths the area per stall space will be some 25% greater than for the 90° car park (see Section 5.1.1).

A two-bin-wide car park with 90° parking could increase its stall-widths to 3 m and still retain its 6 m-wide-aisles without exceeding the area per car space for a two-bin 45° car park with 2.4-m-wide stalls and 3.6 m-wide aisles. Low angled parking should only be employed where the available site width precludes the use of wider bins.

7.5. Split-level decks (SLD)

These have been the most popular circulation layout in the UK for multi-level urban car parks. They can be simple to drive around, and, generally, have a good static and dynamic efficiency. The combination of half-storey-height internal ramps and flat bins enables some types to be constructed down to ten stall-widths in length for two-bin layouts and eight stall-widths for multi-bin layouts while still retaining a complete vehicle circulation and re-circulation capability. In large-capacity facilities, rapid inflow and outflow routes can occur. They must, however, be introduced at the design stage if expensive alteration costs are to be avoided. They can be used with any angle of parking, although only right-angled parking can, sensibly, be used in conjunction with a two-way-flow circulation pattern. Normally, the decks are constructed level and only incorporate drainage falls, in which case storey heights are dictated by the slope and length of the cross-ramps.

When part, or all, of a traffic aisle is made to slope along its length, storey heights can be increased and/or the gradient of the ramps can be reduced. In this manner, a split-level layout can be gradually modified to become another circulation type. The point at which the transition from one type to another occurs can be assumed to be where the slope of the cross-ramps reduces to 5% (one in 20) enabling them to conform to the requirements of the Building Regulations for pedestrian use (K1 Section 2 clause 2.1).

Historically, because of their construction simplicity, circulation efficiency, and ability to be constructed on small sites, an inherent defect created by poor access for pedestrians between split levels tended to be ignored. If they needed to cross over to an adjacent bin, pedestrians were expected to mingle with traffic on the steep vehicle ramps. Sometimes pathways were introduced down the ramp sides, but mostly, pedestrian considerations were ignored in the search to produce the most economical building in a highly competitive market.

Gradually, car park operators and designers began to rectify this defect by introducing dedicated pedestrian ramps and/or stairs between the split-levels. This, however, reduced static efficiency and increased costs. It was not a statutory requirement and in a competitive market many car parks continued to be constructed without the benefit of this improvement.

Current regulations relating to the maximum allowable gradient for pedestrian access between adjacent bins, and the desire for enhanced security, supervision across the decks and the development of other, superior, layout types has rendered the split level layout a less attractive proposition than it has been in the past.

Circulation layouts

Figure SLD 1a One-way flow with an included rapid outflow route

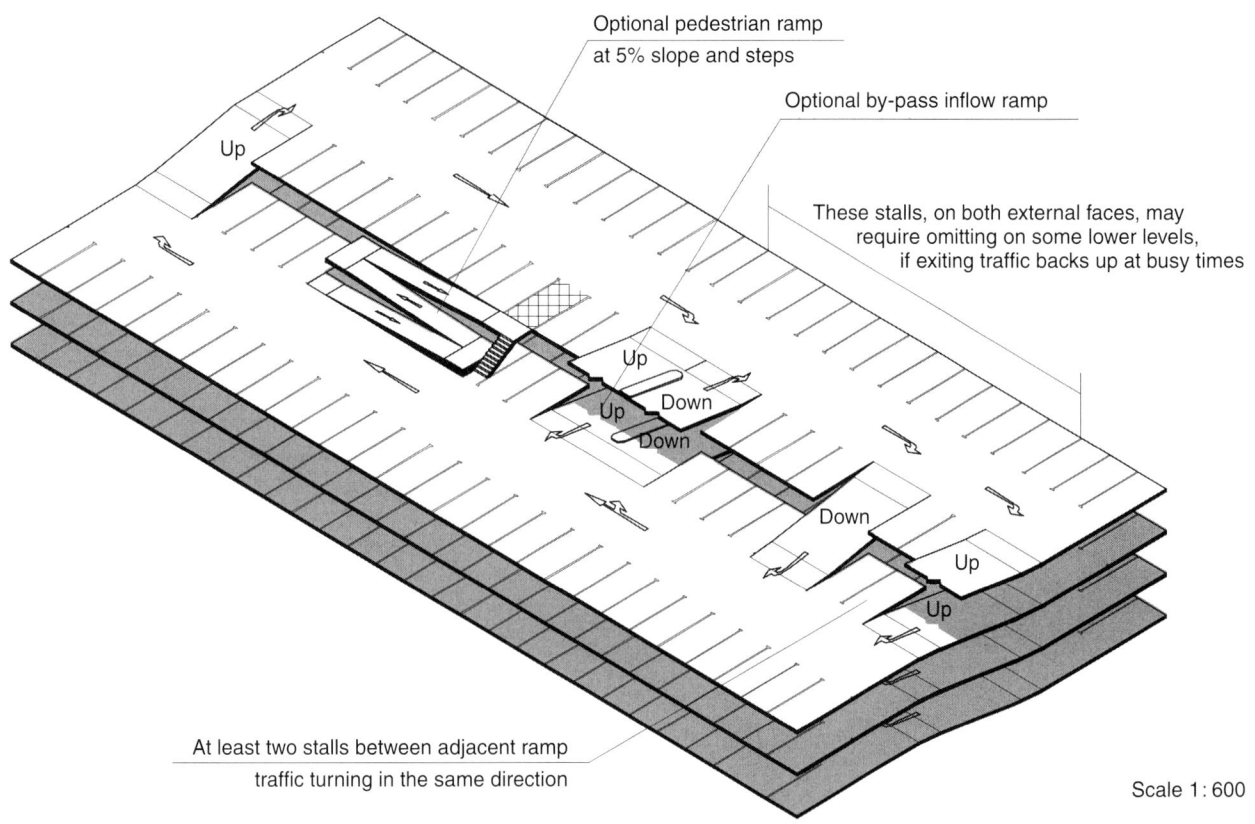

Intermediate levels

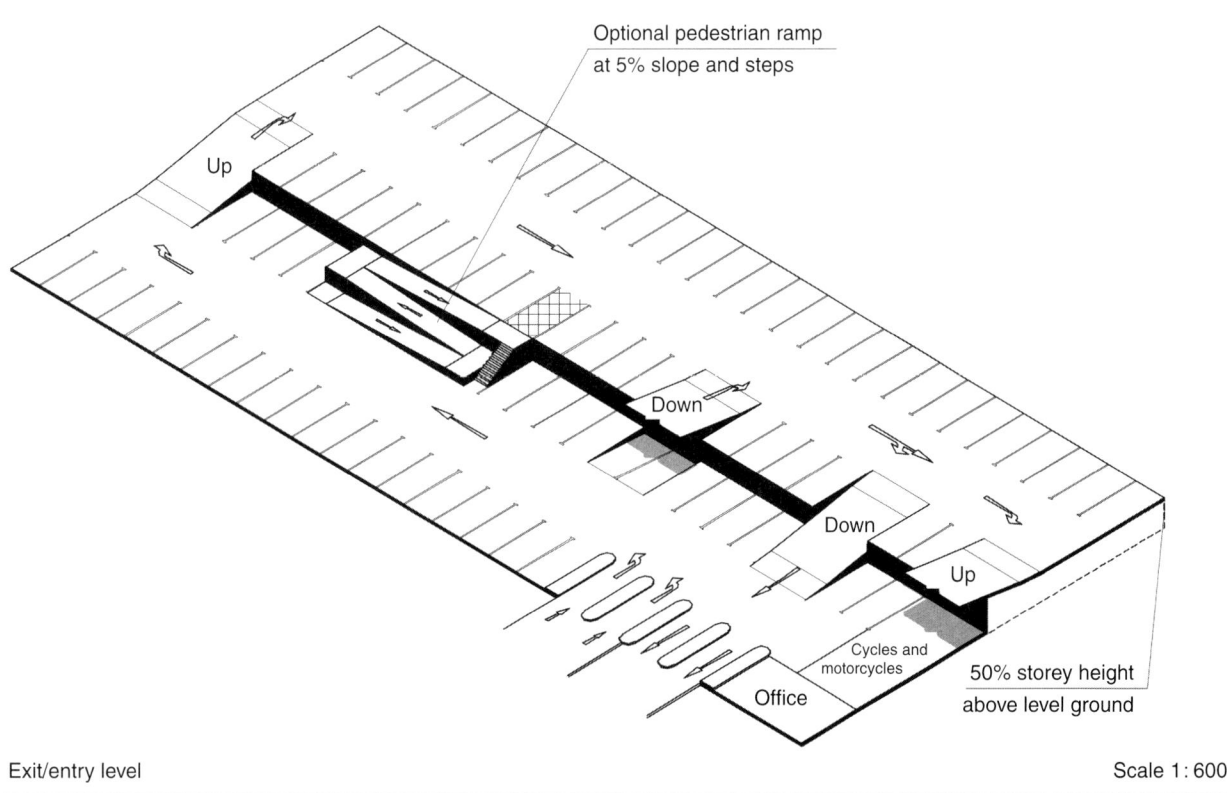

Exit/entry level

Figure SLD 1b One-way flow with an included rapid outflow route

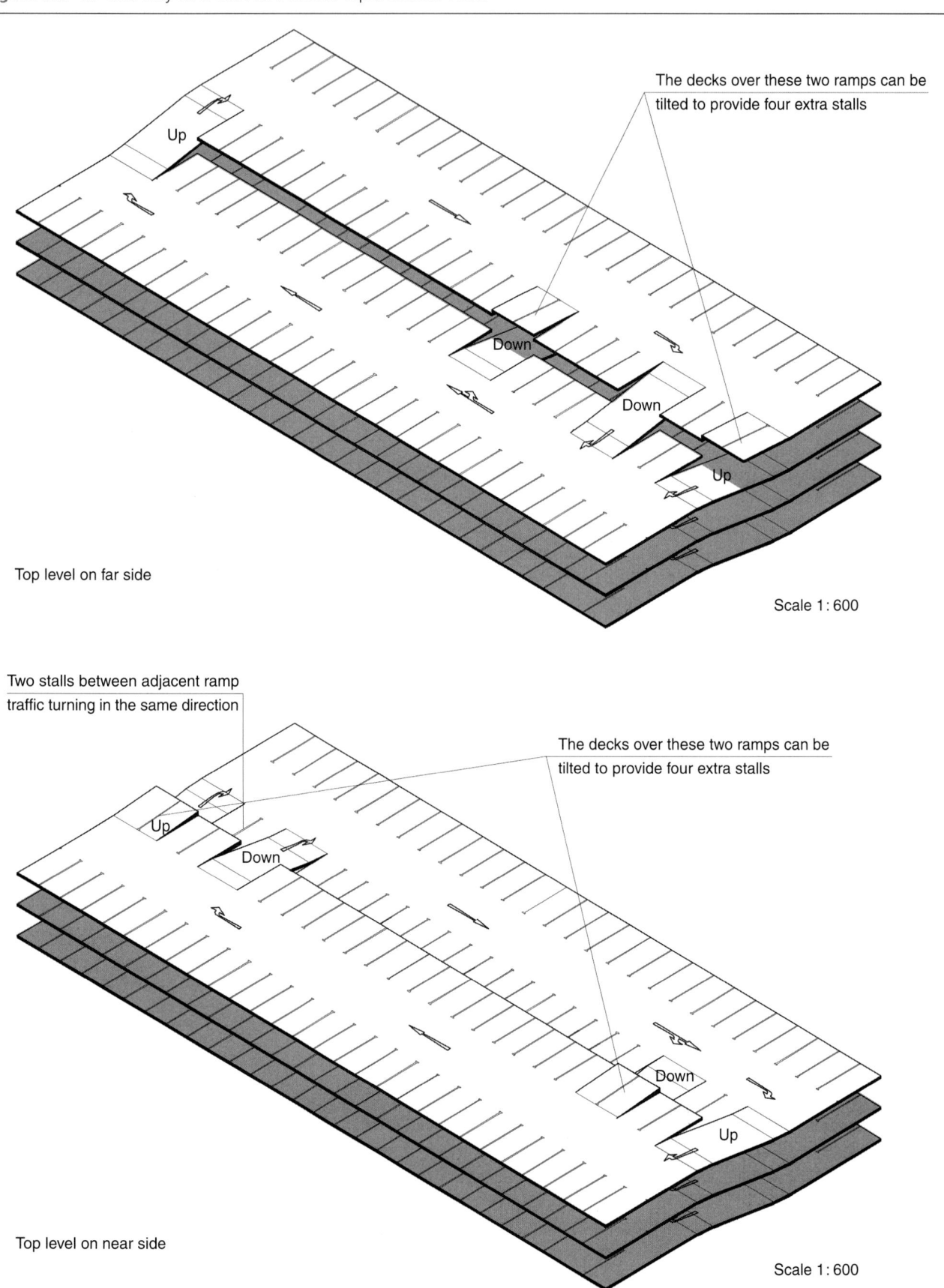

SLD 1 (Figures SLD 1a and 1b)

ADVANTAGES
- The stalls are all located on the main inflow route.
- There is a rapid outflow route.
- All turns are in the same direction.

DISADVANTAGES
- Both flows combine on the outflow route, a condition that can result in traffic congestion at peak times.
- Seven stalls per storey height will be lost if a pedestrian ramp is required between the split levels.

COMMENTS
- This was the basic layout pattern for the 1920s original building in France.
- The layout shows the preferred entry/exit location for an urban centre building where the main access tower occurs on either of the flank walls and the vehicle control barriers must be located within the perimeter. SLD 2 shows another solution.
- If the site slopes and the entry and exit are on the upper level, the lower level can be separated from the upper decks and used for private and/or contract parking.
- When the main access tower occurs at the ends of the traffic aisles, the pedestrian ramp between the split-levels can be omitted with a consequent increase of seven stalls per storey height.
- In Cat. 1 or 2 layouts a capacity of 400 stalls is about the limit before restrictions onto the highway and/or exit control problems result in traffic backing up through the lower decks. In clear-span structures it is possible to route inflow traffic around the exit route with the loss of 20 stalls per storey height, but if internal columns form part of the structure this may be difficult to achieve. Alternatively, a ramp can be introduced at the design stage to by-pass the waiting exit traffic on the lower one or two decks, at a loss of only four stalls per storey height.
- An even number of decks produces the better circulation solution, as the 'far-side' top deck drawing shows. An uneven number of decks results in the top decks having a long 'dead end' requiring another ramp to be constructed, as the 'nearside drawing' shows.
- In busy facilities, when ramps land side by side on the same deck, adjacent driver conflict can occur. It is best to leave at least two stall-widths between them. In small or Cat. 3 or 4 layouts, this is not such a serious problem.
- Tilting the decks over the two top ramps can add another four stalls to the total capacity.
- This layout type is largely obsolescent, at the time of writing. There are, however, many that are still operating successfully but not as efficiently as other more recent designs.
- It is not a recommended layout type and is shown mainly for its historic value.

STATIC EFFICIENCY

As drawn, the number of spaces is 96 and the area per car space, at 21.840 m^2, can be deemed 'Good'.

OTHER LAYOUTS

For larger-capacity layouts an SLD 2 is dynamically superior and more flexible in its design. A VCM 1 is also a superior layout without the need for a pedestrian ramp, and the flat deck areas also render it more user-friendly and economical to construct.

Figure SLD 2a One-way-flow with an excluded rapid outflow route

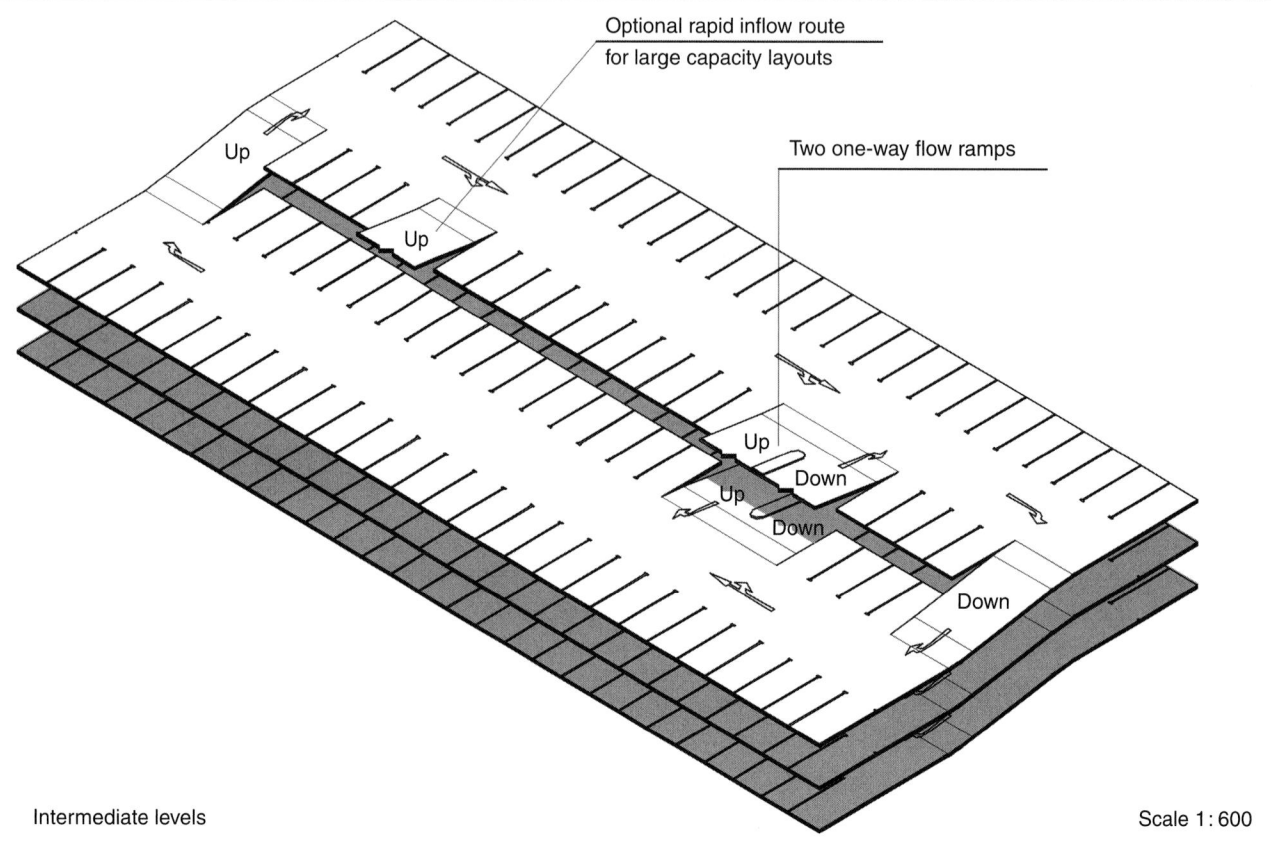

Intermediate levels

Scale 1:600

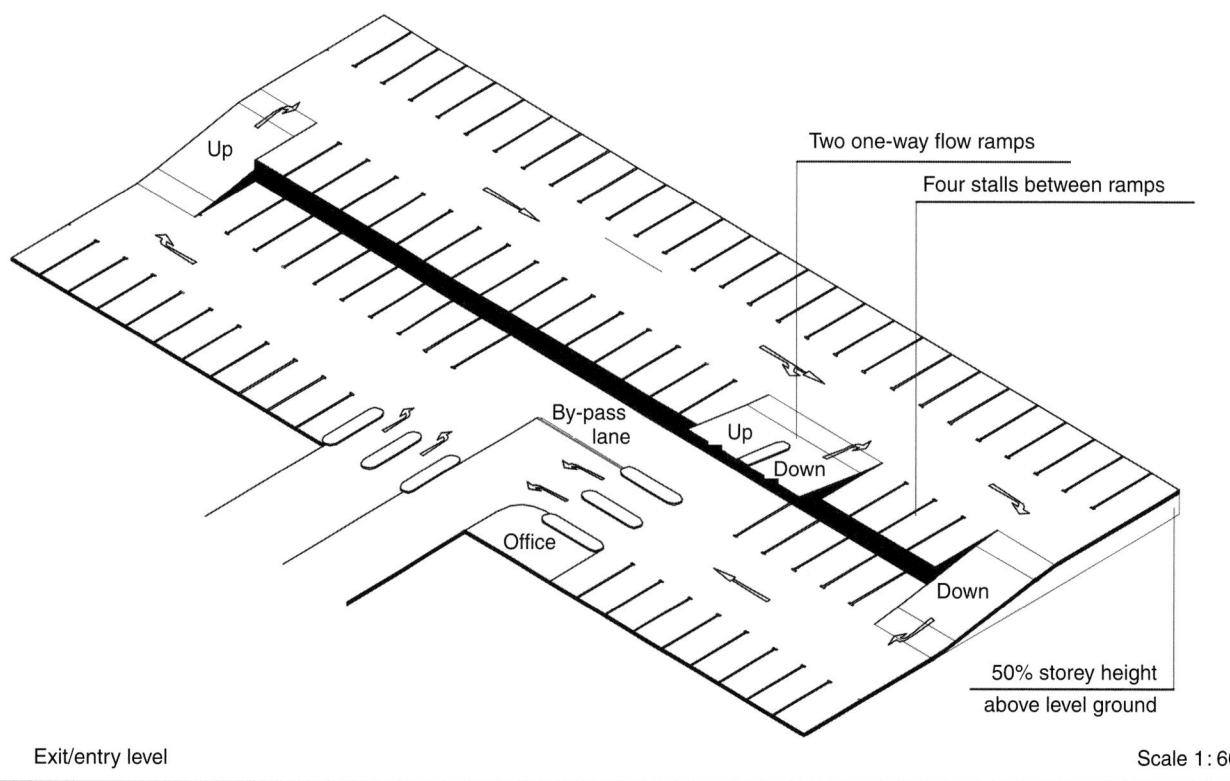

Exit/entry level

Scale 1:600

Circulation layouts

Figure SLD 2b One-way-flow with an excluded rapid outflow route

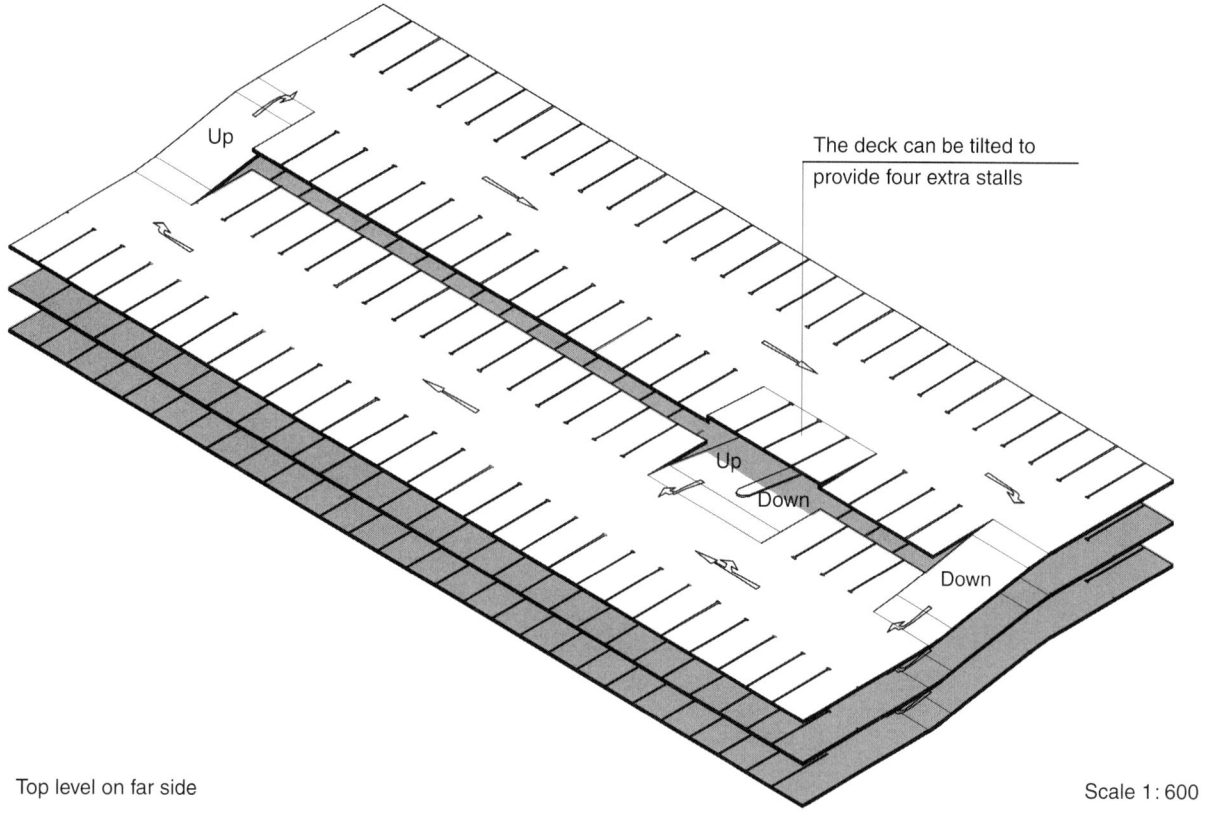

Top level on far side

Scale 1:600

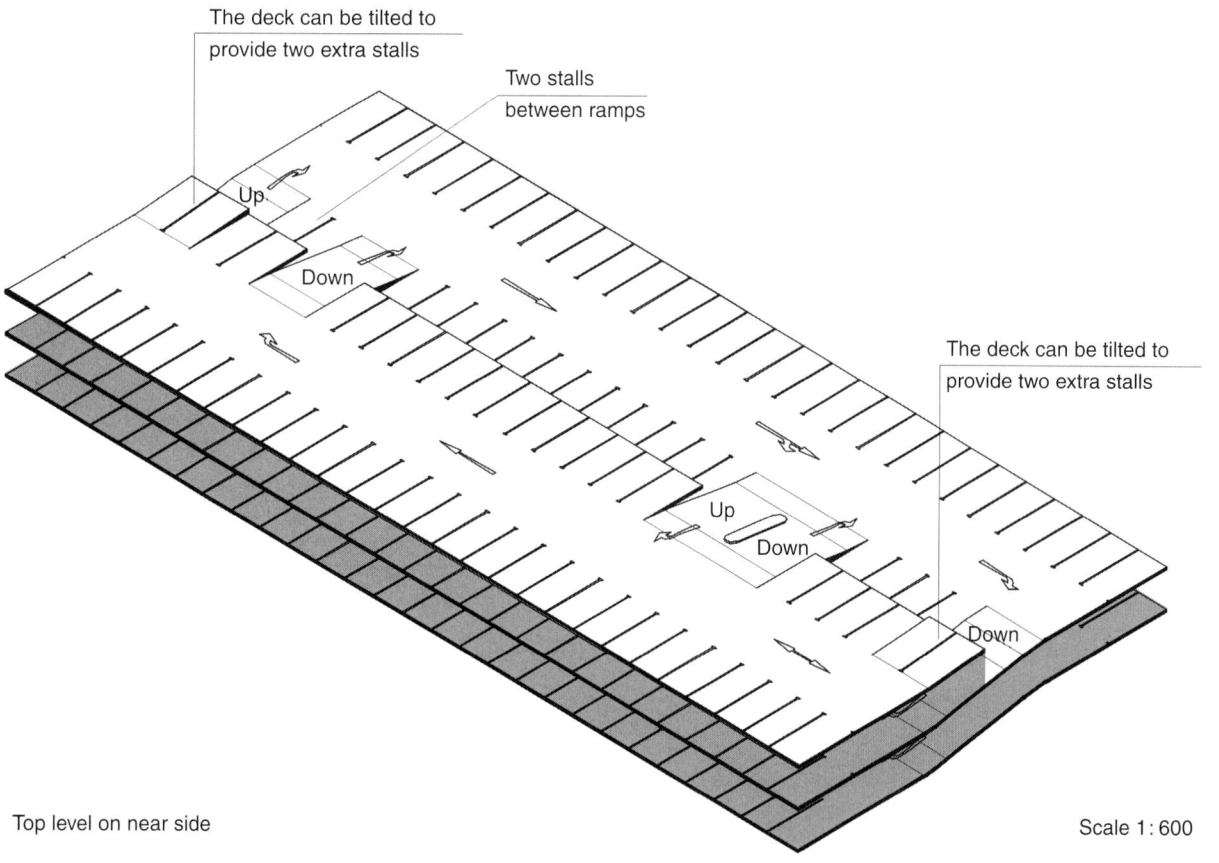

Top level on near side

Scale 1:600

SLD 2 (Figures SLD 2a and 2b)

ADVANTAGES
- It has a rapid outflow route.
- All turns are in the same direction with no turn more than 90°.
- It has a simple re-circulation capability.
- The separation of the flow routes reduces the likelihood of internal traffic congestion.
- Where applicable, the internal ramp width can be reduced, at a saving of two stalls per storey height.

DISADVANTAGE

There is no acceptable pedestrian access between adjacent decks unless ramps and stairs are added, losing seven spaces per storey.

COMMENTS
- This layout type was an improvement made in the mid-1960s to cure circulation problems that were beginning to occur in the larger-capacity SLD 1 car parks that were being built.
- It also shows an alternative entry and exit location. Rather than have a left turn onto the exit barriers, it is preferable to locate them internally and to have drivers turn left on leaving.
- The problem with many exit locations is that once traffic passes the last 'up' ramp, there is no re-circulation capability, and although a long final exit route is good for reducing potential traffic congestion, it prevents drivers from searching that aisle. Therefore, if private or contract spaces are not required to use those stalls, a by-pass route should be considered.
- When larger-capacity facilities are being considered (600+ stalls) a rapid inflow route is worthy of consideration on one or more of the lower suspended decks. By-passing full or congested lower levels enables drivers to rapidly reach the emptier upper levels, and with this in mind the stall capacity for a Cat. 1 or 2 layout can be in the order of 1000 spaces.
- As with SLD 1, a smooth traffic-flow on the top decks relies on an even number of split-levels, as shown on the 'far side' drawing. When there are an odd number of decks, and in order to prevent a long 'dead-end', a 'down' ramp should be introduced, as shown on the 'near side' drawing.
- For Cat. 1 or 2 car parks of less than 400-stalls capacity, consideration could be given to reducing the width of the internal double ramp from four stalls to three.
- Sloping the aisles at either or both ends, as shown in the VCM 1 layout drawing, eliminates the need for internal ramps and stairs for pedestrians between the split levels.
- It is a little out of date and even when its inadequacies have been attended to it still does not have the open parking space of other more recently developed layouts.

STATIC EFFICIENCY
- Given the choice of layout, a two-bin-width facility will always be more statically efficient than one with three or more bins.
- As drawn, the number of stalls is 96, producing an area per stall of 21.840 m² that can be deemed 'good'.

OTHER LAYOUTS

When the site slopes are in the order of 10%, this layout type can be useful in keeping down the costs at ground level, but otherwise it cannot compete with a VCM 1 design.

Circulation layouts

Figure SLD 3a One-way flow with scissors-type ramps

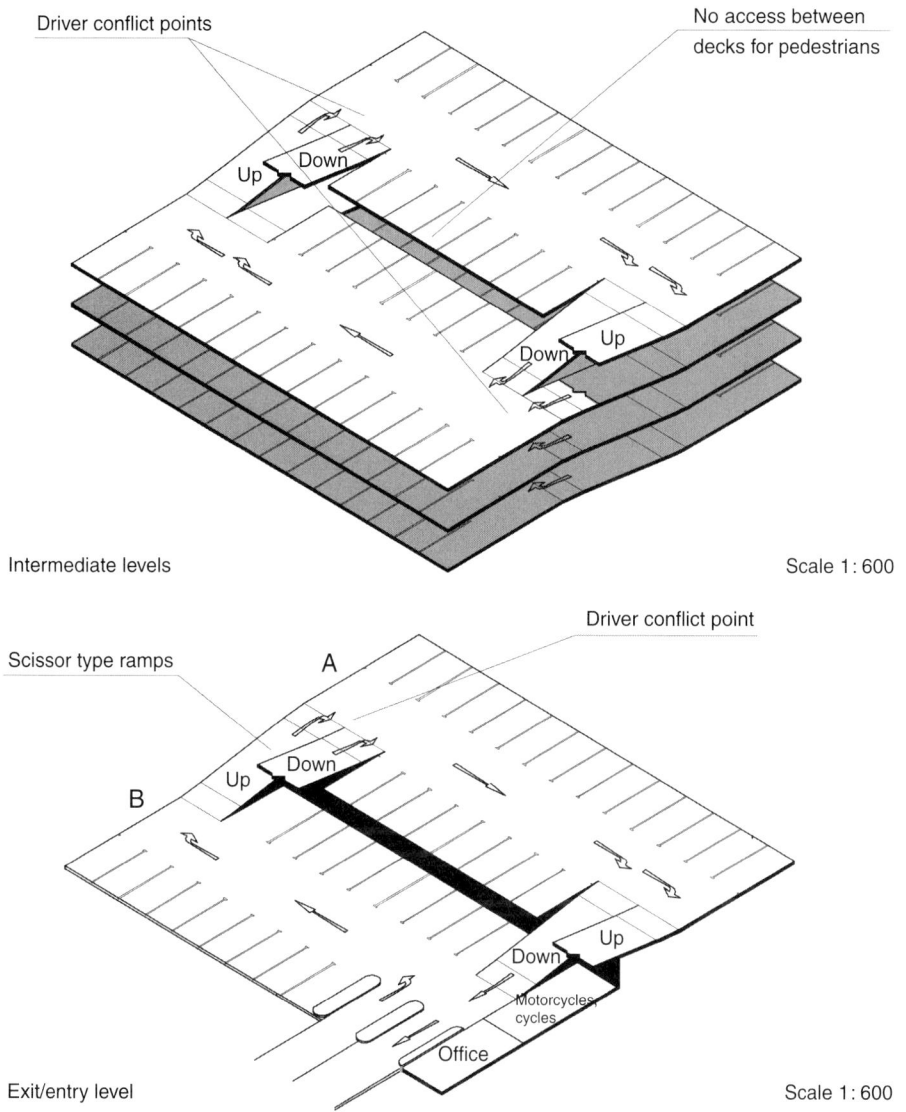

Figure SLD 3b One-way flow with scissors-type ramps

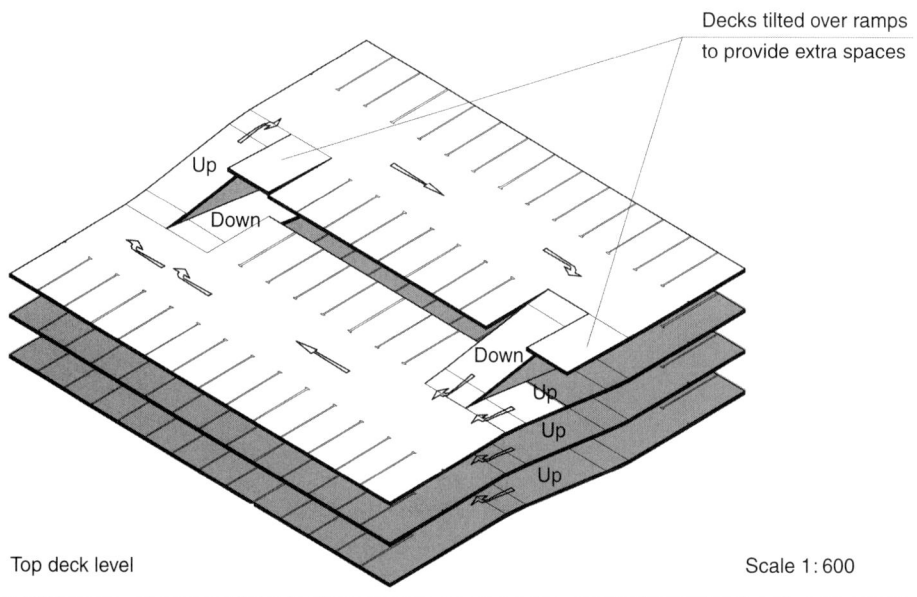

SLD 3 (Figures SLD 3a and 3b)

ADVANTAGES
- It can be used for layouts down to 24 m in length.
- All turns are in the same direction with no turn greater than 90°.
- It has a simple re-circulation capability.

DISADVANTAGES
- The flow routes are combined.
- The scissors-type ramps in landing side by side can create turning problems for drivers; however, in Cat. 3 or 4, or in small Cat 1 or 2 layouts, this may not create too much of a problem.
- The ramps are narrow, as the scratches down the sides of existing ramps testify.
- Pedestrian access between adjacent decks may prove difficult to achieve unless the access tower is located on a gable end.

COMMENTS
- It has been used mainly for small, Cat. 3 or 4 use, where the site area is small, the parking need is great and where the narrow ramps will be tolerated by drivers. However, many public car parks were also constructed to this pattern in the 1960s and 1970s. They are unpopular, especially with drivers of SDV-sized vehicles.
- It should only be used for small (up to 200 stalls maximum), non-public layouts.
- The location of the entry/exit barriers is the most efficient for fiscal control, but if there is no control, such as for a staff car park, then any sensible location will do.

STATIC EFFICIENCY
- Only a minimum of 12 stall spaces per storey height are required to effect the vertical circulation. This is deemed 'good'. The area per stall varies significantly, depending on the overall length of the building. At its shortest length of circulation (24 m), it is 26.75 m^2 per stall and at 68.5 m it reduces to 21.06 m^2. However, it is not recommended that the maximum length should exceed 16 stalls, at which length the layout can take on the form of an SLD 1 building.

OTHER LAYOUTS
- A SLD 4 is to be preferred, especially in a 'tidal' flow situation where bin widths can remain as for one-way flow traffic.
- A VCM 3 can also be used to advantage and has a significant area of flat deck that benefits pedestrians.

Figure SLD 4 Two-way flow aisles and ramps

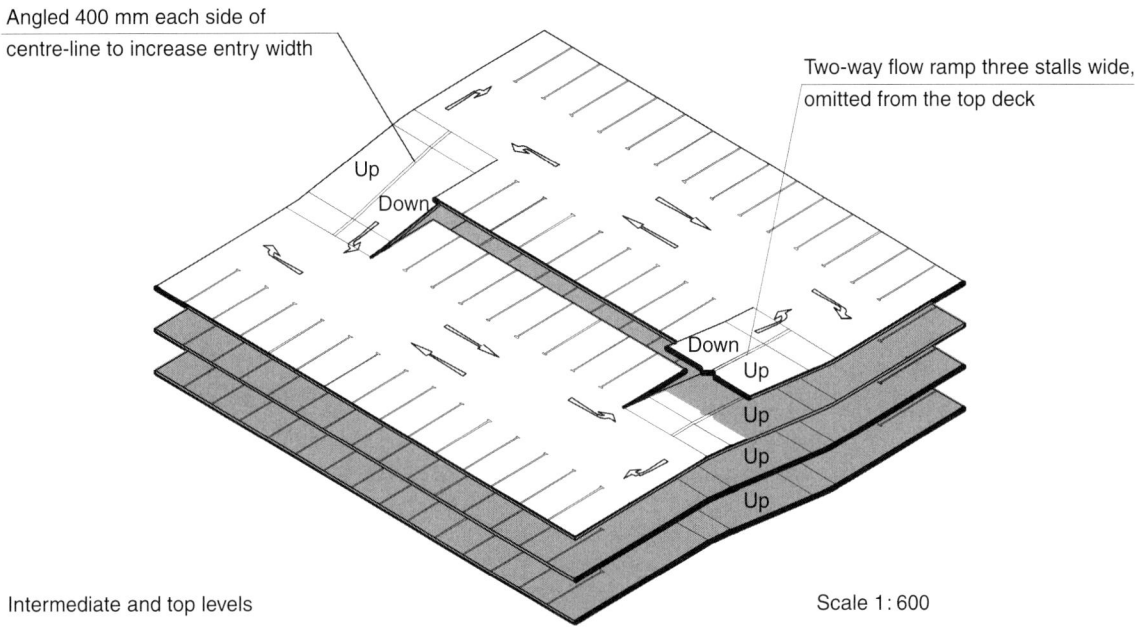

Intermediate and top levels   Scale 1:600

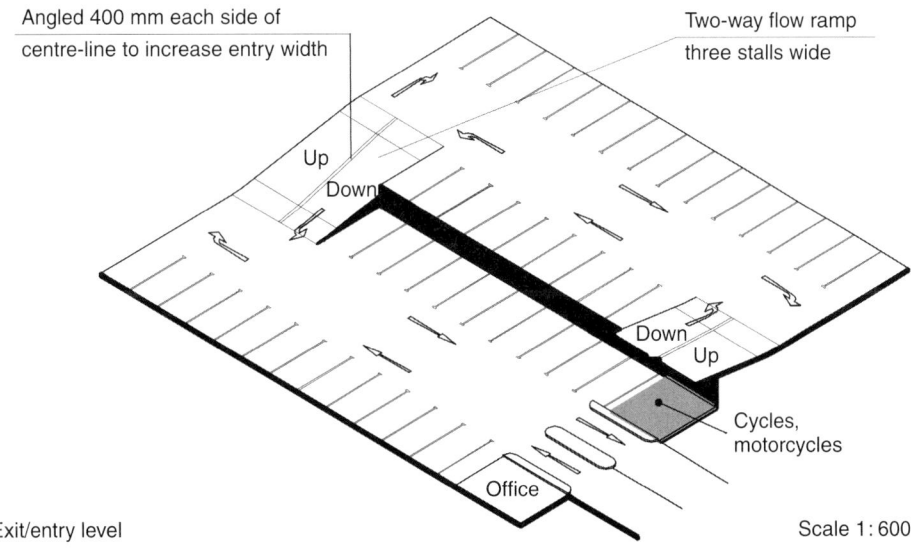

Exit/entry level   Scale 1:600

Circulation layouts

SLD 4 (Figures SLD 4)

ADVANTAGES

- All stalls are located directly off the main inflow route.
- It can be used in layouts down to 24 m in length.

DISADVANTAGES

- All stalls are located directly off the outflow route.
- Re-circulation involves turning around on a traffic aisle.
- There is limited or no access for pedestrians between split-levels.

COMMENTS

- This is a two-way-flow version of the SLD 3 layout but without the narrow ramps and potential driver conflict.
- It is a layout more suitable for Cats. 3 or 4 purposes, especially 'tidal' where the full width of the ramp can be used, a.m. and p.m.
- For Cat. 4 use, or where very light use is anticipated, the width of the aisles can be reduced to that for one-way-flow traffic; however, the client's approval should be obtained first.
- Suitable for all parking categories up to about 300 stalls, the lack of a rapid outflow route and the two-way flow preclude this layout from serious consideration for larger Cats. 1 and 2 purposes.
- It is a rather dated layout and can only be recommended for current use as a staff- or small hotel-type use less than 12 stall-widths in length, although there are many longer than that operating in the public sector at the present time.

STATIC EFFICIENCY

- It has a static efficiency similar to that for a SLD 3 layout.

OTHER LAYOUTS

- Down to a length of 14 stall-widths, a normal VCM 3 layout is superior in dynamic efficiency and 'user-friendly' features and when using a minimum-length VCM module it can be reduced to 12 stall-widths.

Circulation layouts

Figure SD 3a Double helix end connected with one-way-flow on the central aisle

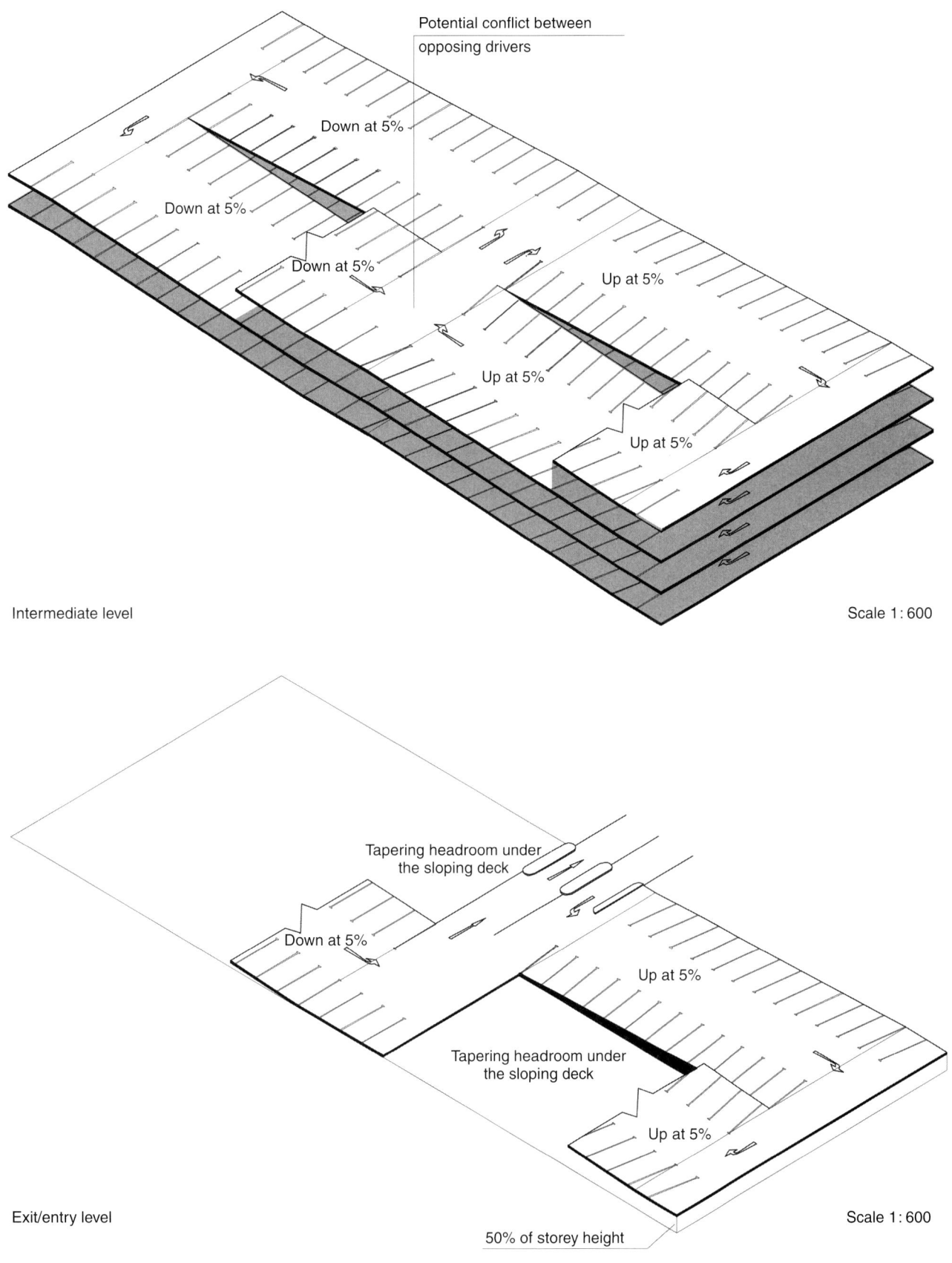

Figure SD 3b Double helix end connected with one-way-flow on the central aisle

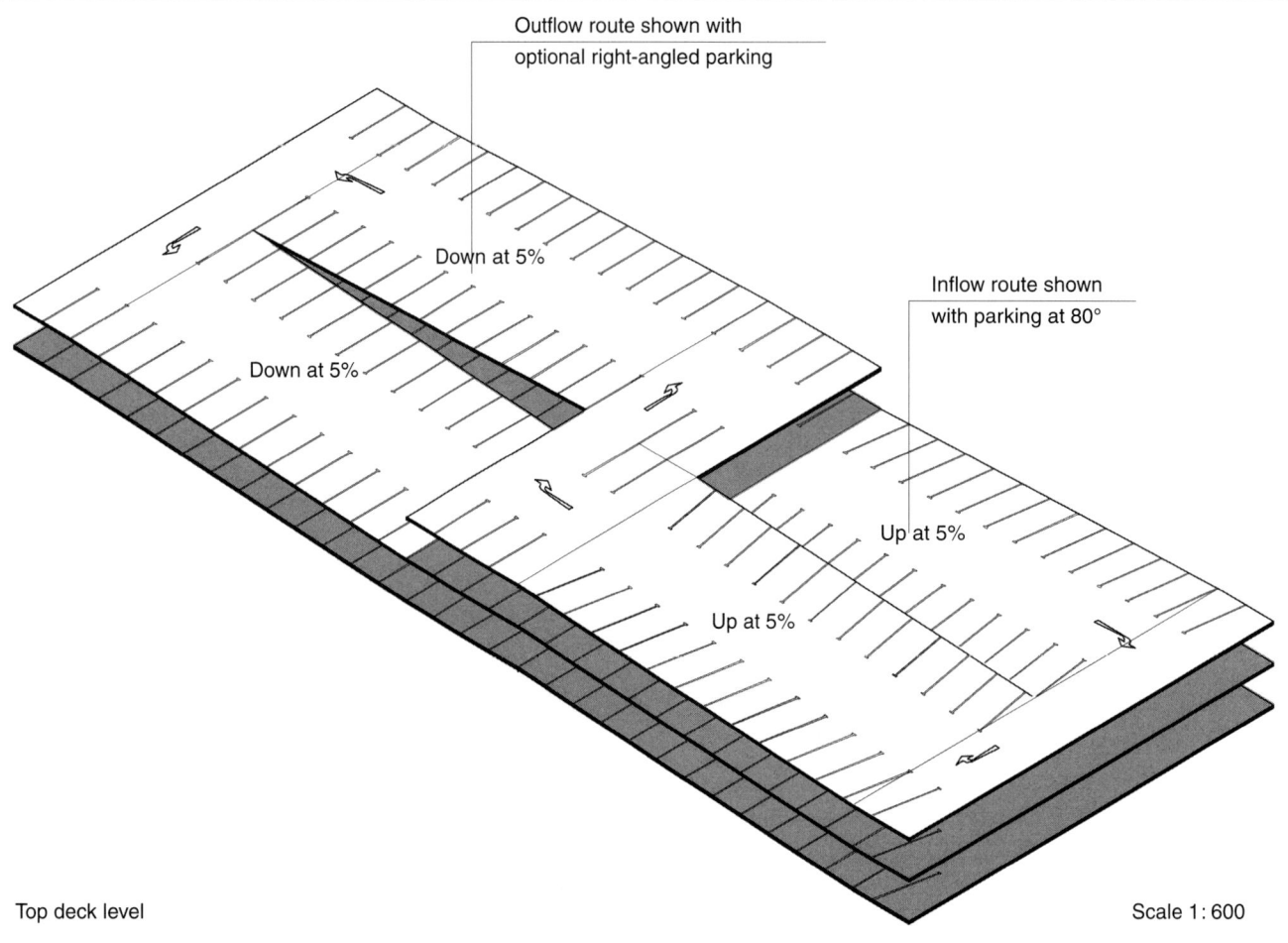

Top deck level

Scale 1 : 600

SD 3 (Figures SD 3a and 3b)

ADVANTAGES

- Flat cross-ways are located at each end and in the middle.
- There is a simple circulation and re-circulation capability.
- Storey heights can be varied by the ramp length.
- Disabled driver stalls, offices etc. can be located separately below the tapering areas under the sloping decks.

DISADVANTAGES

- Only 50% of the parking capacity is covered by the inflow route; the remainder must be accessed from the downward-flowing outflow route, requiring an extended search pattern in order to pass all of the stalls.
- If used for Cat. 1 or 2 purposes, Variable Message Signing (VMS) is desirable to prevent extended search patterns from causing traffic congestion at busy times.
- Drivers on both flow routes are in confrontation when they both turn onto the central cross-way at the same time.
- There is no rapid inflow or outflow route capability.
- It cannot be constructed under a length of 74.4 metres (31 stall-widths).

COMMENTS

- This is a popular layout form in the USA, where it is often seen linked to an adjacent multi-storey office block and used for staff parking. In such buildings, the traffic-flows are mainly 'tidal' and the use of the central cross-way for both flow routes is not an important factor.
- With this layout the elevations display a continuous stepped rise (or fall) on each long side, and the central cross-way incorporates a one way-flow route.
- If more than a single entry/exit control point is required then the flat central section must be widened.
- One way of passing all of the stalls without 'double driving' through the aisles is to climb to the top deck on the inflow route and come back down on the outflow route. That, however, will not be a very popular search pattern.
- This is a one-way-flow system and to prevent drivers from turning down-hill, against the design intentions, angled parking should be used for the inflow route, at least.
- Since down-hill is a natural direction for exiting traffic, 90° parking stalls can be used, as shown, although consistency could well make them the same for all of the aisles.
- They are not a popular layout form in the UK and are not recommended for Cats. 1 or 2 uses.

STATIC EFFICIENCY

- At the minimum length of 72.4 m, the area per car space produced is 21.1 m^2. This can be deemed to be 'Good'.

OTHER LAYOUTS

- VCM 1 and 2 layouts, incorporating rapid outflow routes and extended length inflow routes, together with good deck pedestrian access, render them more acceptable design forms for all categories of use.
- An SLD is also worthy of consideration provided that cross-deck access is not a necessary design feature.

Car Park Designers' Handbook

Figure SD 4a Double helix end connected with two-way-flow on the central aisle

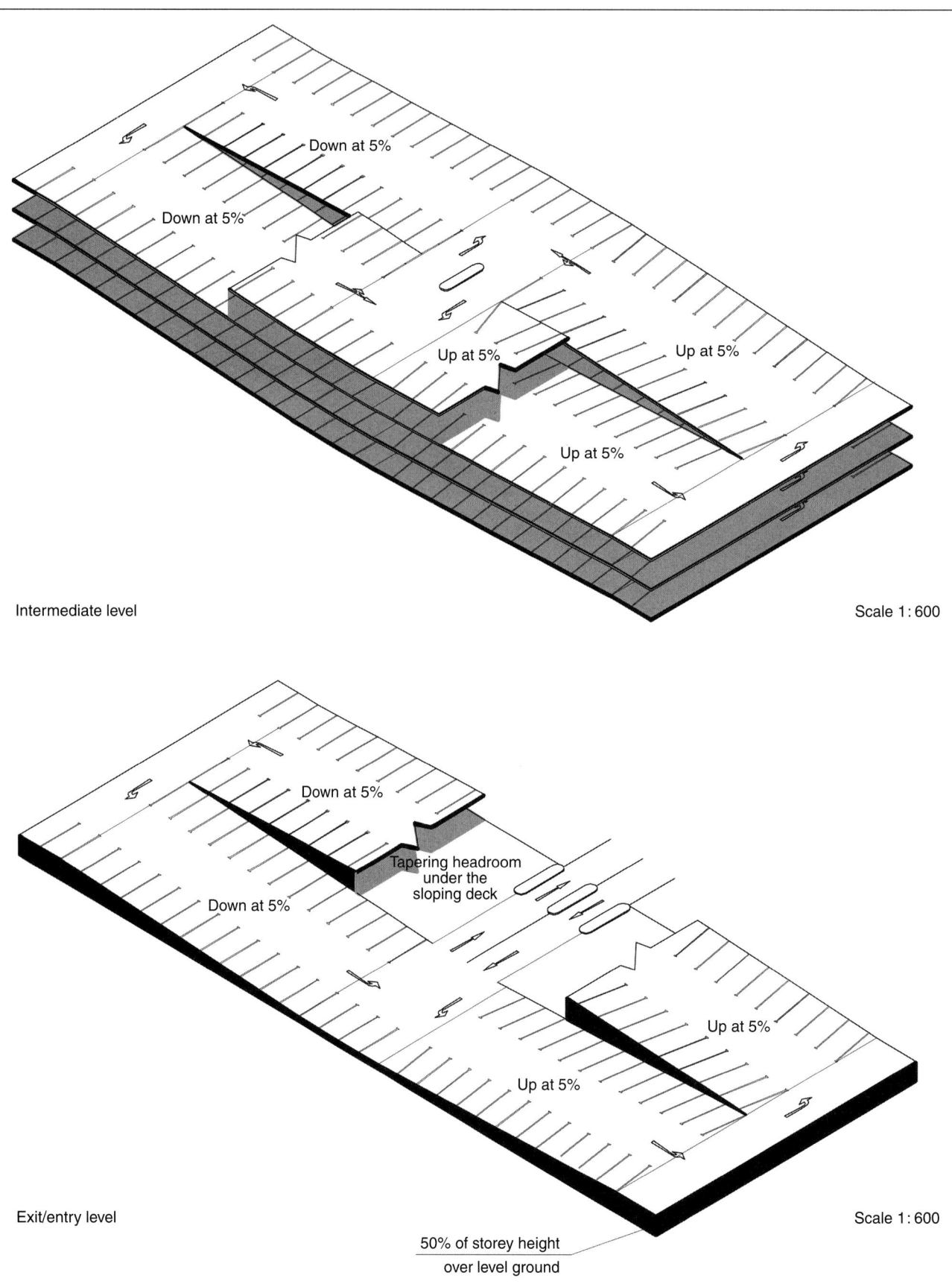

Figure SD 4b Double helix end connected with two-way-flow on the central aisle

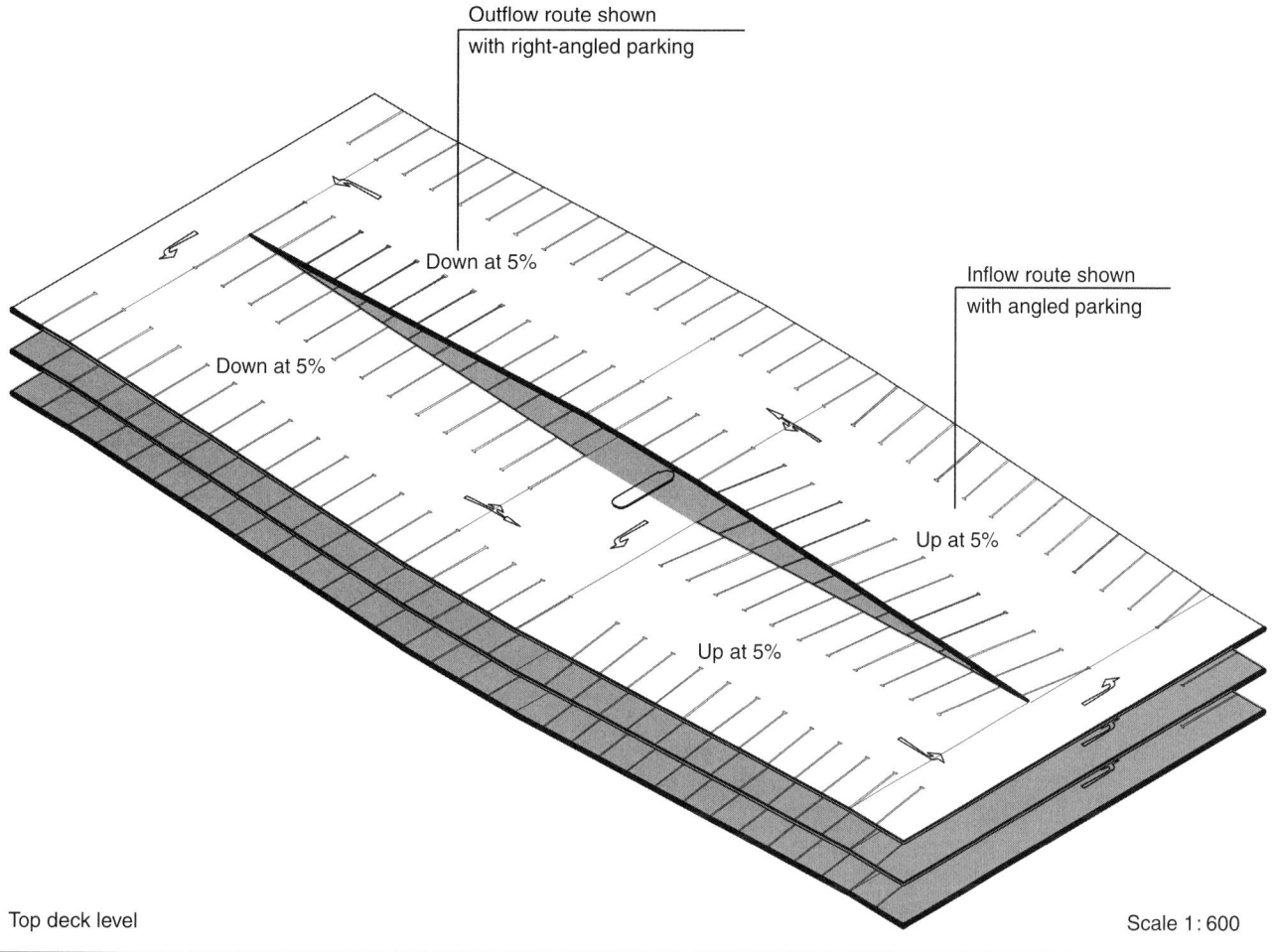

SD 4 (Figures SD 4a and 4b)
ADVANTAGES
- Flat cross-ways are located at each end and in the middle.
- Simple circulation and re-circulation capability.
- Storey heights can be varied by the ramp length.
- Disabled driver stalls, offices, etc. can be located separately below the tapering areas under the sloping decks.
- All turns are made in the same direction.

DISADVANTAGES
- Only 50% of the parking capacity is covered by the inflow route; the remainder must be accessed from the downward-flowing outflow route, requiring an extended search pattern in order to pass all of the stalls.
- If used for Cat. 1 or 2 purposes, Variable Message Signing (VMS) is desirable to prevent extended search patterns from causing traffic congestion at busy times.
- Drivers on both flow routes are in confrontation when they both turn onto the central cross-way at the same time.
- There is no rapid inflow or outflow route capability.
- It cannot be constructed under a length of 76.8 metres (32 stall-widths).

COMMENTS
- This is also a popular layout form in the USA, often seen linked to an adjacent multi-storey office block and used for staff parking. In such buildings the traffic-flows are mainly 'tidal' and the use of the central cross-way for both flow routes is not an important factor.
- With this layout the elevations display a more balanced appearance on each long side, either anhedral or dihedral, and the central cross-way incorporates a two-way-flow route.
- A centrally located entry and exit is the only sensible configuration for these layout types.
- One way of passing all of the stalls without 'double driving' through the aisles, is to climb to the top deck on the inflow route and come back down on the outflow route. That, however, will not be a very popular search pattern.
- This is a one-way-flow system and to prevent drivers from turning down-hill, against the design intentions, angled parking should be used for the inflow route, at least.
- Since down-hill is a natural direction for exiting traffic, 90° parking stalls can be used, as shown, although consistency could well make them the same for all of the aisles.
- They are not a popular layout form in the UK and are not recommended for Cats. 1 or 2 uses.

STATIC EFFICIENCY
- At the minimum length of 76.8 m, the area per car space produced is 21.4 m^2. This can be deemed to be 'Good'.

OTHER LAYOUTS
- VCM 1 and 2 layouts incorporating rapid outflow routes and extended-length inflow routes, together with good deck pedestrian access, render them more acceptable design forms for all categories of use.
- An SLD is also worthy of consideration provided that cross-deck access is not a necessary design feature.

Circulation layouts

Figure SD 5 Double helixes with one-way-flows

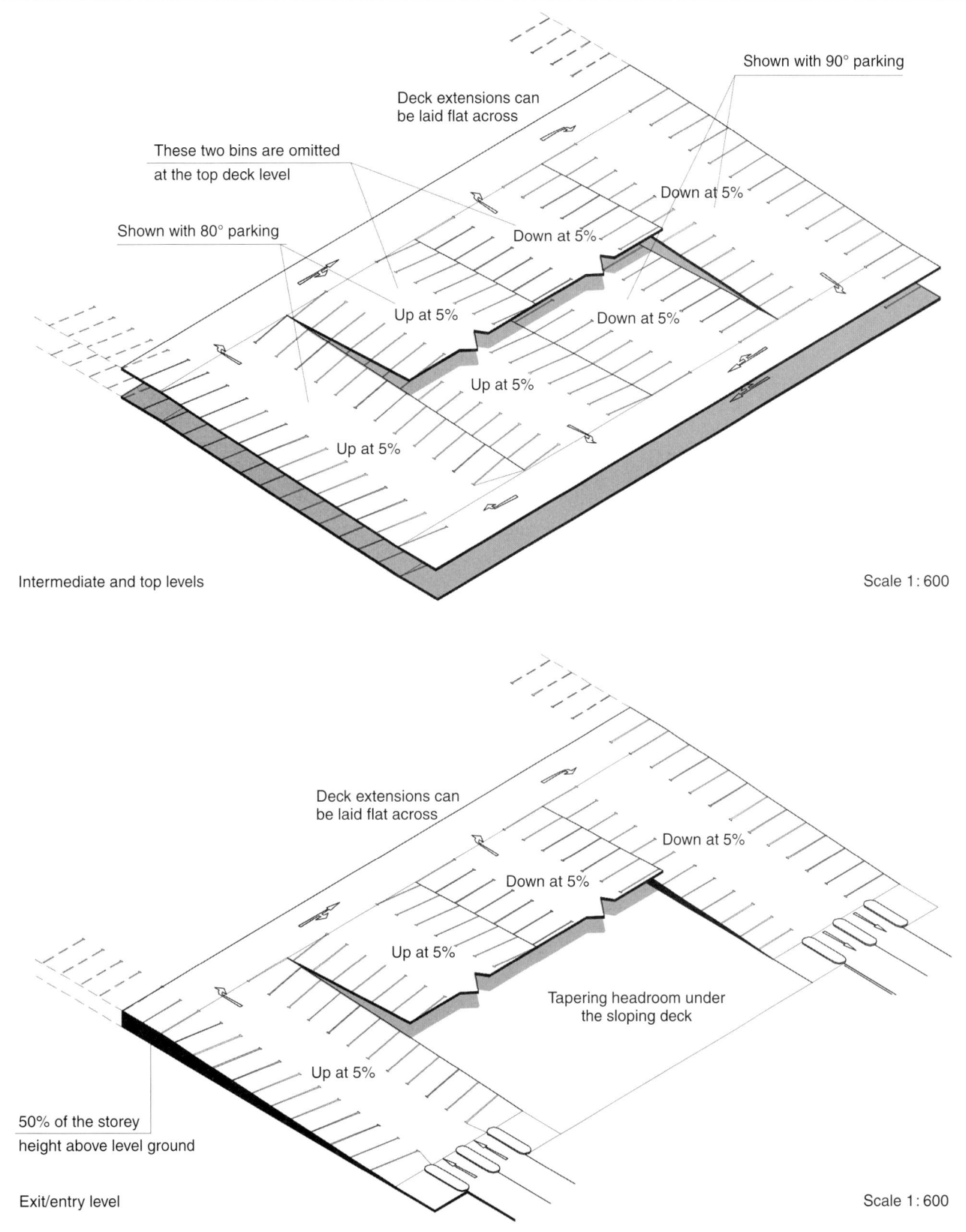

SD 5 (Figure SD 5)
ADVANTAGES
- Flat pedestrian-accessible cross-ways are located on the aisle ends.
- It has a simple circulation and re-circulation capability.
- Storey heights can be varied by the ramp length.
- All turns are made in the same direction.
- Both flow routes can be semi-rapid regardless of the overall dimensions of the decks.
- Disabled driver stalls, offices, etc. can be located separately below the tapering areas under the sloping decks.

DISADVANTAGES
- Only 50% of the parking capacity for the minimum dimension layout is covered by the inflow route; the remainder must be accessed from the outflow route, requiring an extended search pattern in order to pass all of the stalls. However, when used in a larger-deck capacity layout, this percentage diminishes.
- If used for Cat. 1 or 2 purposes, Variable Message Signing (VMS) is desirable to prevent extended search patterns from causing traffic congestion at busy times.
- It cannot be constructed under a length of 38.4 m (16 stall-widths).

COMMENTS
- Variations on this type are a popular layout form in the USA, sometimes seen linked to an adjacent multi-storey office block, but used for all parking categories.
- The entry/exit points can also be located side by side in the central aisles by reversing the layout.
- Although shown at its minimum, any increase in building length should be laid flat across with significant improvements in user satisfaction.
- One way of passing all of the stalls without 'double driving' through some of the aisles, is to climb to the top deck on the inflow route and come back down on the outflow route. It will not however, be a very popular search pattern.
- In the aisles area one-way flow system, and to prevent drivers from turning down-hill against the design intentions, angled parking should be used for the inflow route, at least.
- Since down-hill is a natural direction for exiting traffic, 90° parking stalls can be used, as shown, although consistency could well make them the same for all of the aisles. The central aisle parking, however, must be at right angles.
- They are not used to any great extent in the UK.

STATIC EFFICIENCY
- At the minimum length of 38.4 m, the area per car space produced is 23.04 m^2. This can be deemed to be 'Average'.

OTHER LAYOUTS
- Ninety-six stalls occur on the sloping deck elements compared with only 24 on a similar VCM-type layout, and the 'unassisted' stall search route is greatly extended.
- With less flat-deck areas in contact with the ground, it is also more costly to construct the lowest deck levels.

7.7. Flat and sloping deck layouts (FSD)

Although one side of a pair of decks is laid flat, it has no contact with the other sloping side and for the length of the sloping part (24 stall-widths) pedestrians are unable to cross over. Sometimes it has been seen that the sloping element runs the length of the building regardless of the lack of cross deck access.

The similarity between this series of layouts and the SD series is quite close, and for this fact only two layouts have been shown that have been constructed and have some justification in being shown.

The practical advantage of these layouts over those featured in the SD series is mainly architectural, in that the parking decks can be laid horizontally on three sides of the building in a two-bin configuration and on all four sides of buildings that are three or more bins wide; apart from that, there is little advantage in their use.

The length of the deck, raising a complete storey height, renders these layouts suitable only for buildings of a length greater than about 72 m (30 stall-widths).

Rapid inflow or outflow routes are not a practical proposition.

End-connected flow routes require a minimum building length of 128.8 m to operate successfully. They are not considered to be a viable construction form for UK use and for this reason they have not been featured.

They are quite popular in the USA, where they are used mainly for 'main terminal' and staff-type parking in multiple-bin layouts, linked to large office buildings.

They are rare in the UK.

Figures FSD 1 and FSD 2 Single helix with two-way flow

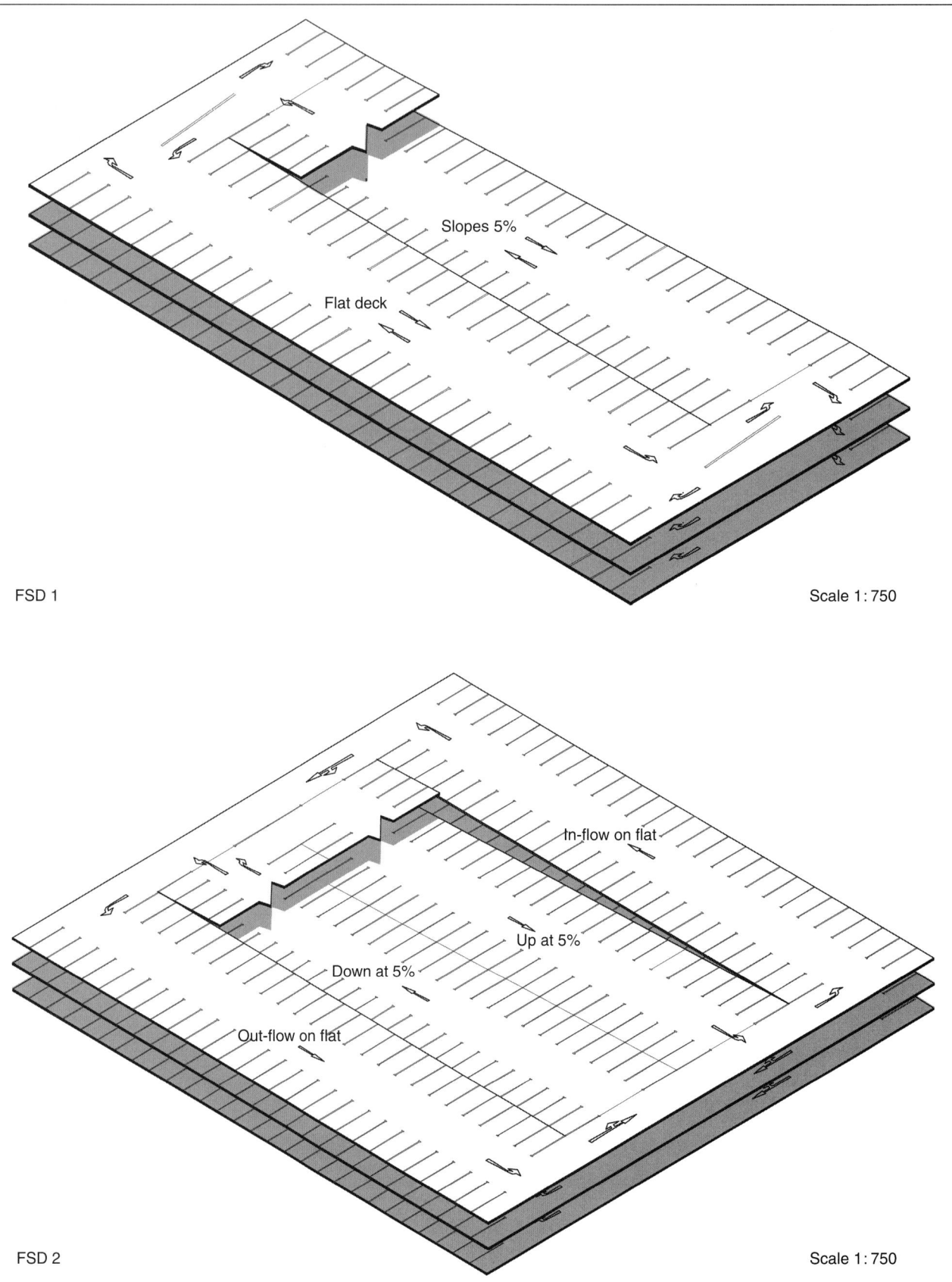

FSD 1

FSD 2

FSD 1 and 2 (Figures FSD 1 and 2)
These are similar to the SD series but in raising the sloping parking decks a complete storey height in a straight line, their minimum lengths are doubled.

In the first edition of this book, FSD 2 is simply an extended length SD 2 layout; FSD 3, 4, 6 and 7 are so similar that they have been considered unnecessary to include in this second edition. FSD 5 has been included simply to show that the layout can be constructed flat on all four sides.

ADVANTAGE
Storey heights can be varied by the length of the 5% sloping parking deck, if sufficient site space is available.

DISADVANTAGES
Added to those of the similar SD layouts are the following disadvantages.

- When raising 3 m in one straight slope, the minimum building length is about 30 stall-widths (72 m) and if the storey height requirement is greater, then this will be extended commensurately.
- Cross-aisle access for pedestrians is restricted to the ends of the long aisles.
- Rapid-flow routes are not a practical proposition.

COMMENTS
- They can be seen in large-capacity layouts in the USA, especially large terminal buildings, but there are very few in the UK.
- There are other layouts that are superior both dynamically and statically, and in user-friendly features.
- They are not a layout pattern that can be recommended, except in the very few cases where there may be no alternative.

STATIC EFFICIENCY
A minimum deck-length contains 108 stalls and produces an area per car space of 22.1 m^2. This can be deemed 'Good'.

OTHER LAYOUTS
- Any of the SD series can be constructed in half of the length with the remaining length constructed flat across the bins.
- An SLD 4 layout has similar cross-bin access problems to FSD 1, but the bins are constructed flat.
- VCM 3 also has a similar circulation pattern to FSD 1, but with more than 66% of the deck area constructed flat across both bins.
- Compared with a minimum sized FSD 5 layout, a VCM 1 or 2 will have more than 80% of the deck area laid flat across the bins, with a choice of rapid flow routes.

7.8. Combined flat and sloping deck layouts with internal cross ramps (VCM, FSDR and WPD)

VCM (Vertical Circulation Module) is a circulation system that uses modules two bins wide by eight or more stalls in length to create a minimum-dimension (for good practice) circulation route.

Stacking one module upon another creates a continuous vertical circulation flow for a structured parking facility, and joining two modules at their ramps (one the mirror image of the other) creates a complete car park circulation system where pedestrians can reach all parts of the deck without negotiating any slope in excess of 5%.

The modules can be a minimum dimension of eight stall-widths in length with a slightly warped or a 5% sloping access-way, or they can be extended in length to ten stalls which renders them more suitable for pre-cast concrete structures. Outside of the modules, the parking decks are flat, resulting in a significant improvement in cross-deck accessibility for pedestrians. The modules can be moved individually to other locations as can be seen in VCM 2. Modifying a module to accept two-way traffic-flows creates a layout that can be constructed down to ten stall-widths (24 m) in length, which still retains pedestrian access to any part of the deck (VCM 3).

A VCM layout is more user-friendly than other internally ramped systems and the 5% slopes provide access to any part of a parking deck, thereby eliminating the need for a separate dedicated pedestrian ramp. This reduces construction costs, enhances static efficiency and increases the user-friendliness. In large capacity parking decks, rapid inflow routes can be introduced at any future time and at minimal cost: it is simply a matter of line painting. One-way flow layouts can be constructed with a minimum dimension of eight stall-widths while still retaining flat access for pedestrians along one long side (see the MD series).

The WPD (Warped Parking Deck) is a circulation system that provides the external appearance of a residential or office building, with all four sides appearing to have flat decks, without slopes or steps. However, internal deck slopes of up to 9% laterally and 5% longitudinally must be tolerated. They were a popular layout form in the 1960s and 1970s, but no longer conform to current requirements. It is thought that some examples of this type of building are still operating.

They cannot be constructed less than 74.4 m in length. Rapid-flow routes are not a practical proposition and pedestrian access between adjacent decks is restricted to the aisle end. It is similar in circulation to a SD 3 layout.

The FSD 3 layout has a single straight sloping deck that raises 50% of a storey height combined with a single cross-ramp to get to another level. It is similar in circulation to a SD 3 layout, but without the central access-way for pedestrians.

Circulation layouts

Figure VCM 1a One-way flow with internal ramps

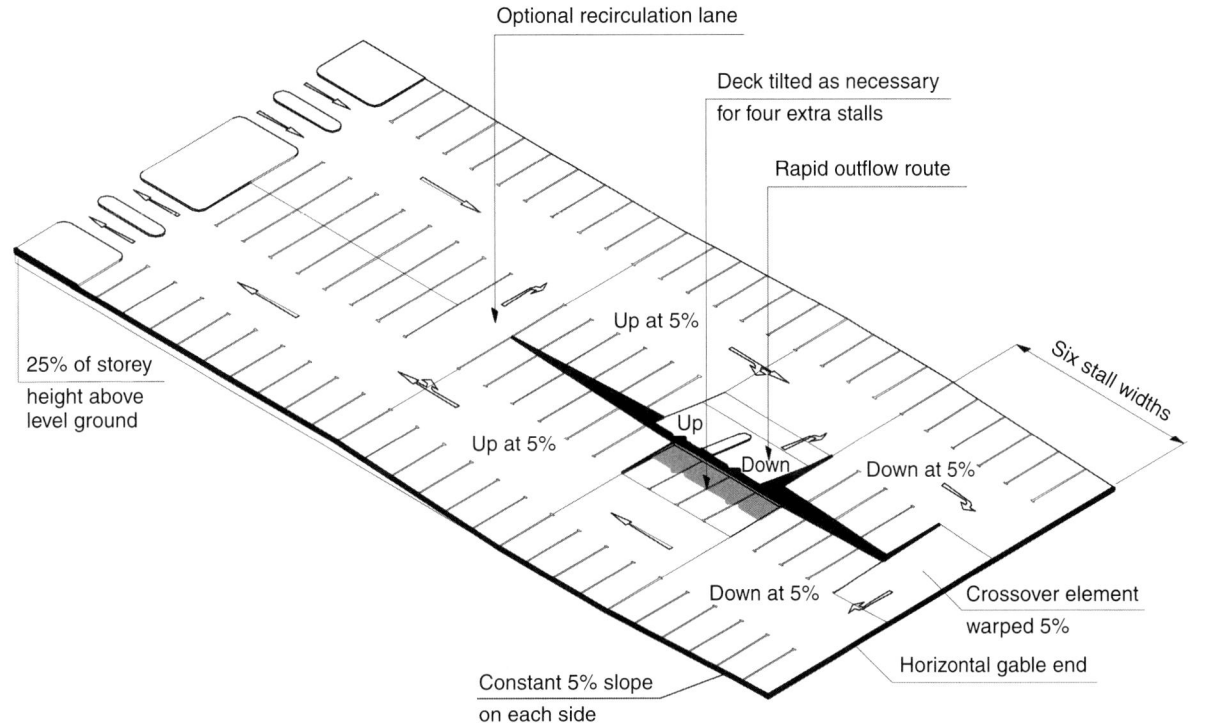

Exit/entry level (with gable end variation)

Scale 1 : 600

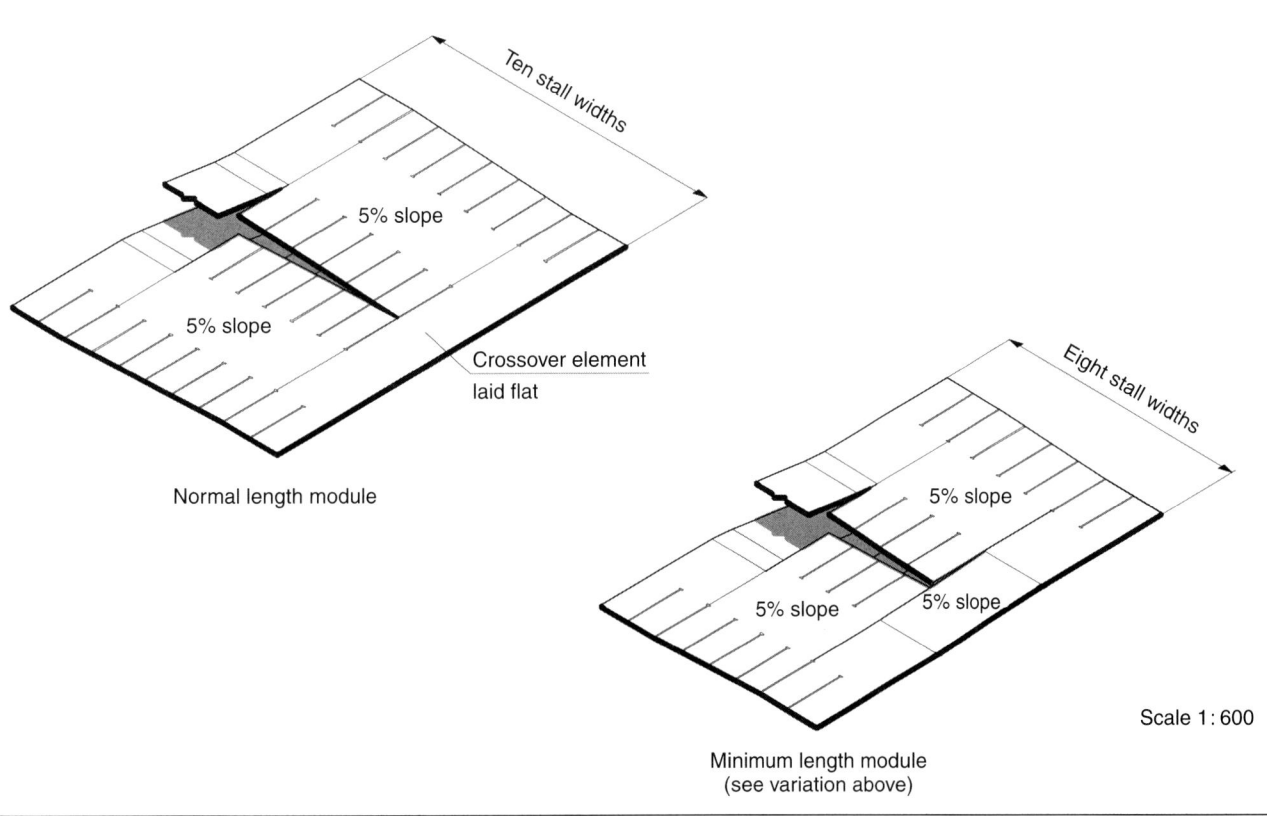

Normal length module

Minimum length module
(see variation above)

Scale 1 : 600

Figure VCM 1b One-way flow with internal ramps

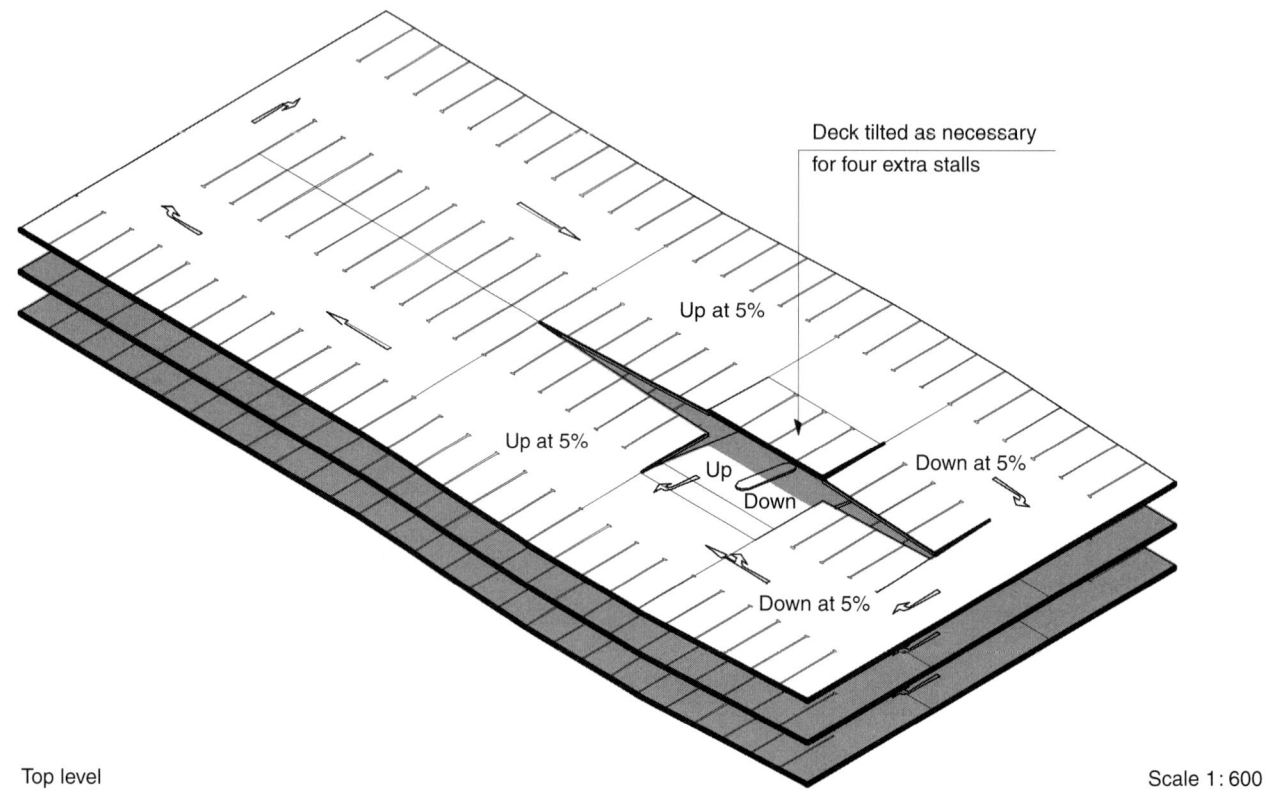

Top level   Scale 1:600

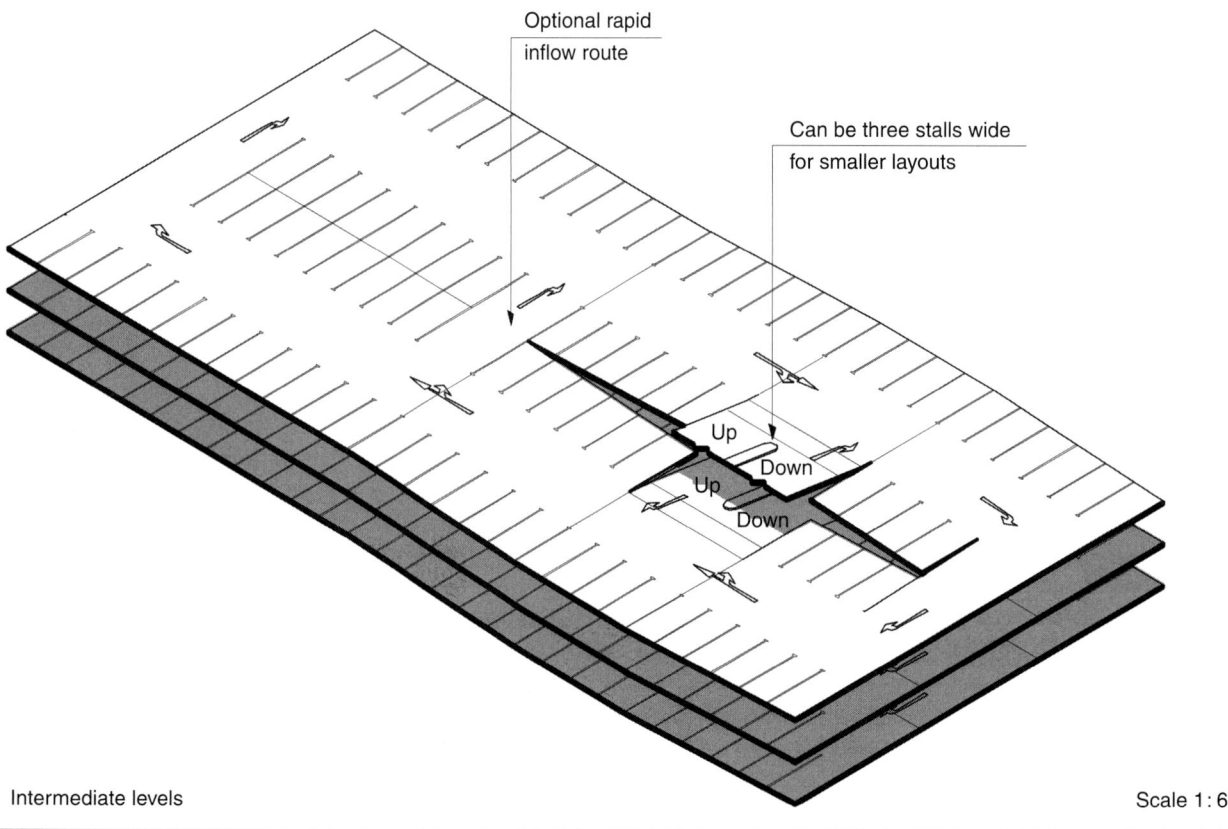

Intermediate levels   Scale 1:600

Circulation layouts

VCM 1 (Figures VCM 1a and 1b)

ADVANTAGES
- It is pedestrian-accessible throughout the decks without any slopes in excess of 5%.
- It has a simple circulation and re-circulation capability.
- Storey heights can be varied by the length of the sloping deck elements.
- All turns are made in the same direction.
- Both flow routes can be made rapid regardless of the overall dimensions of the decks. It is simply a matter of painting lines.
- Only 24 stalls per storey need be passed on the outflow route.

DISADVANTAGES
- It cannot be used in layouts less than 36 m (15 stall-widths) in length.

COMMENTS
- With the entry/exit positions located at the aisle ends, most of the ground-floor deck can be laid directly on prepared ground. Only the relatively small sloping-deck areas need a small amount of 'cut and fill'.
- In the smallest of this layout type, with the cross-ways at the aisle ends, only eight extra stall spaces are required to complete the vertical circulation (six if a combined central ramp is used). As shown, with longer layouts, 12 or 16 will be required; even so, this is quite acceptable.

There are three types of vertical circulation module (VCM), as follows.

- As shown at the end of the long layout, the last eight stalls on each side slope continuously to the gable end producing a 5% warped condition for the cross-way element. In practice it is unnoticeable to drivers.
- When the two end stall-widths are constructed flat, the cross-way slopes 5% as shown in the minimum length module.
- In the normal length module the last two stall-widths are laid flat and the cross-way also is flat.

All of these variations have been used successfully to solve layout design requirements.

- With the short sloping inflow aisle, angled parking has not been found to be necessary.
- Dynamic efficiency is good; it is similar to a SLD 2 layout but is more acceptable to pedestrians.

STATIC EFFICIENCY
- As shown, there are 96 stalls. The area per car space is 21.84 m^2 and can be deemed 'good'.
- Eliminating the cross-way for the rapid inflow route can improve the static efficiency, but may be desirable on one or two of the lower deck levels. A decision can be delayed until dictated by actual circumstances. It is simply a matter of line-painting.

OTHER LAYOUTS
- A sideways-sloping site could render a SLD 1 layout worthy of consideration.
- A VCM 4 layout is also of similar efficiency but cannot be constructed under 32 stall-widths (76.8 m) in length.

Figure VCM 2a One-way-flow with end ramps

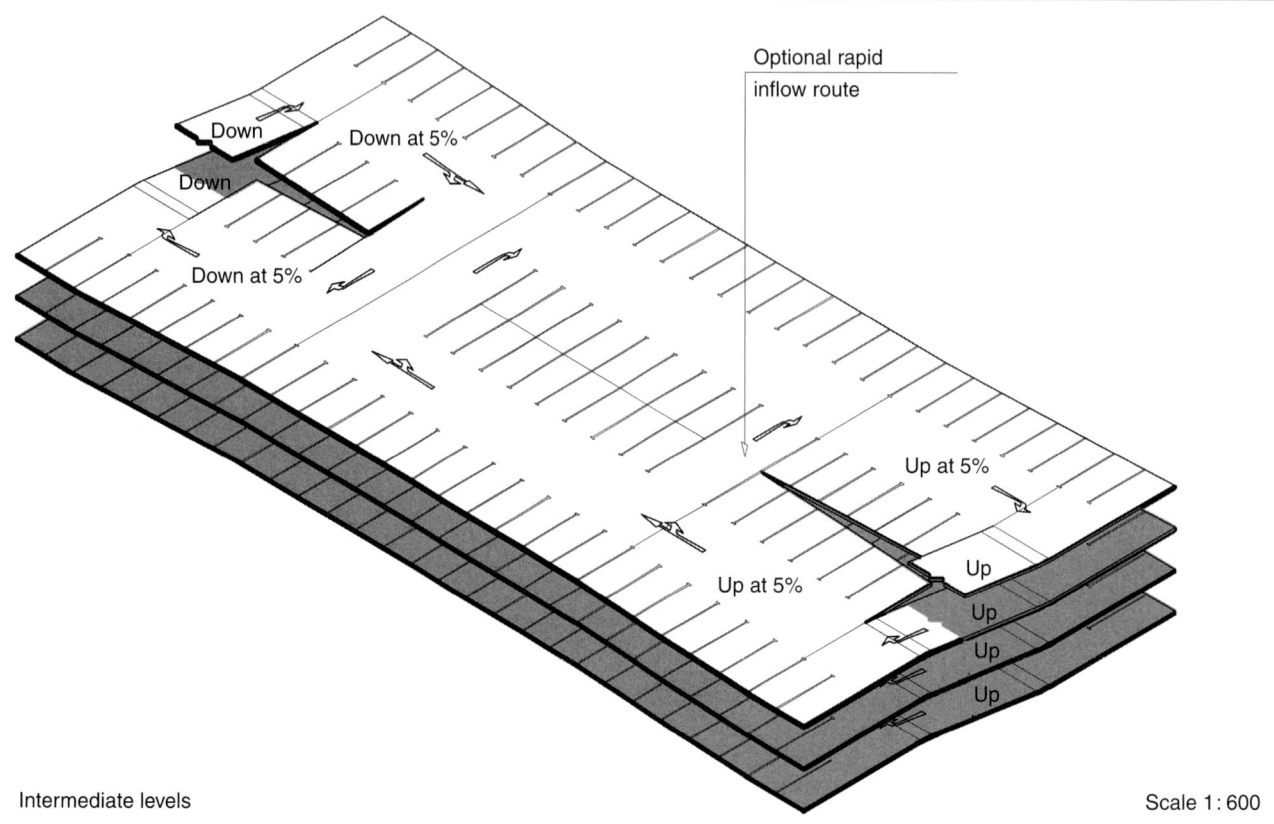

Intermediate levels

Scale 1 : 600

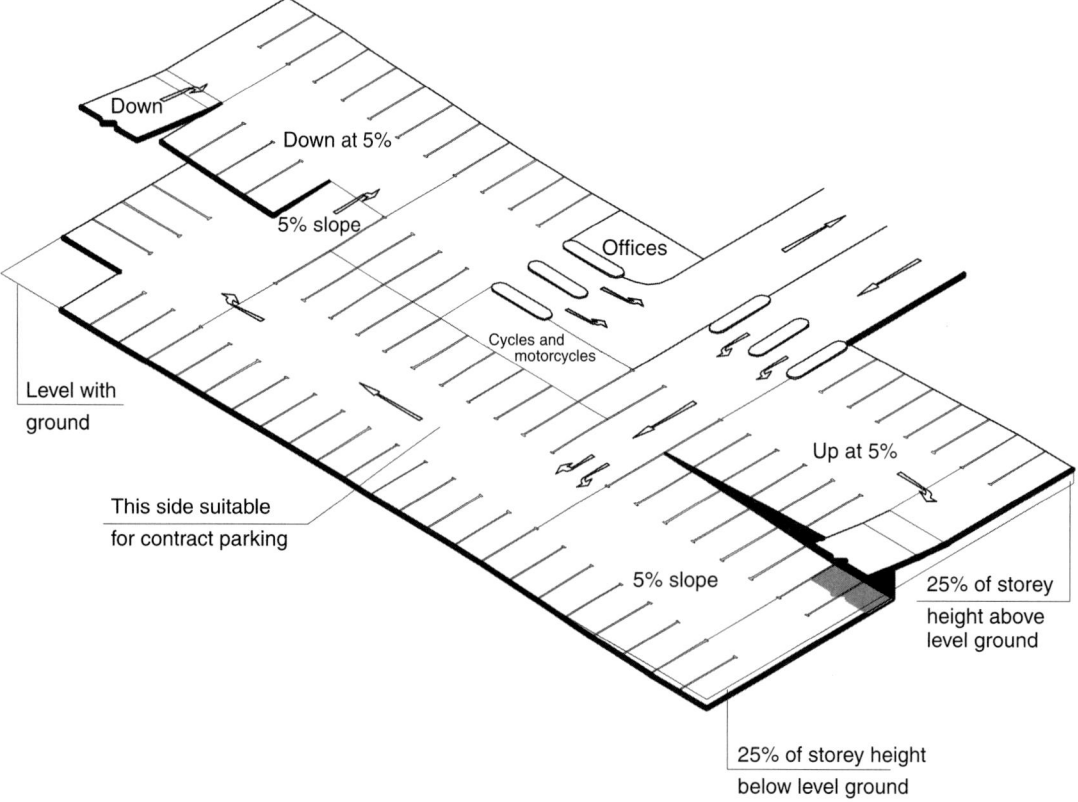

Exit/entry level

Scale 1 : 600

Figure VCM 2b One-way-flow with end ramps

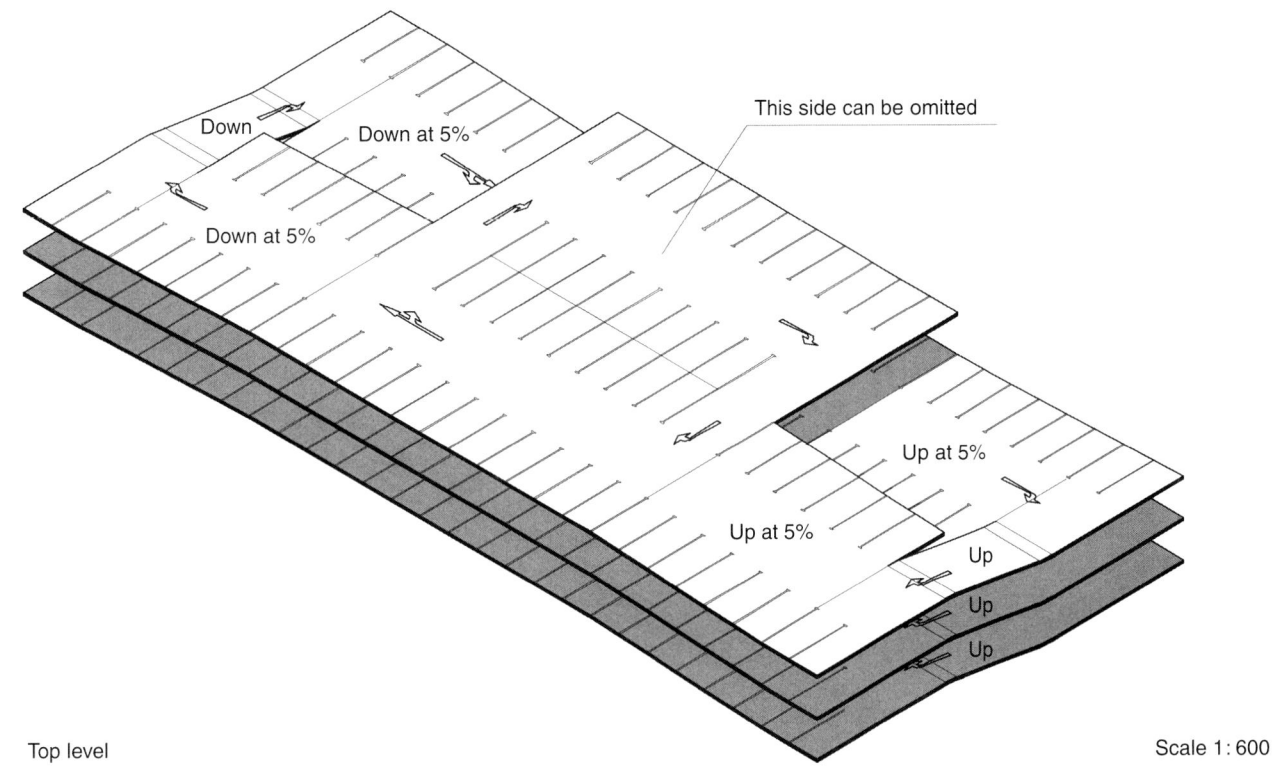

Top level

Scale 1:600

VCM 2 (Figures VCM 2a and 2b)

ADVANTAGES

- As shown, all turns are made in the same direction, but this might alter if it becomes part of a much larger layout.
- It is pedestrian-accessible throughout the decks without encountering slopes in excess of 5%.
- It has a simple circulation and re-circulation capability.
- Storey heights can be varied by the length of the sloping deck elements.
- Both flow-routes can be made rapid, regardless of the overall dimensions of the decks. It is simply a matter of painting lines.
- Only 24 stalls per storey need be passed on the outflow route.

DISADVANTAGE

It is not such a neat solution on the bottom and top decks where there are dead ends or a few lost stalls.

COMMENTS

- Assuming that the ground conditions are flat and horizontal, the entry/exit position on the side of this layout is the most practical location. It has been shown to eliminate the problems associated with left-turning onto a control kiosk, although other fiscal methods may eliminate that problem. This access location can also be used for a VCM 1 layout provided that the flat-deck area is big enough.
- Dependent on the overall building length, there can be a significant area of flat deck in the middle section and although shown as a two-bin layout, the modules can be located at any convenient position in a much larger deck area.
- It is not really suitable for small- to medium-capacity layouts where the VCM 1 layout with its flat ends is more flexible. It does, however, come into its own in large-capacity decks where the modules can be separated and relocated to other positions.
- Only eight stall spaces are required to complete the vertical circulation, regardless of the capacity of the deck.
- With the short, sloping inflow aisle, angled parking has not been found to be necessary.

STATIC EFFICIENCY

- It is similar to VCM 1, but if 'dead ends' are to be avoided at ground level and the top deck, it might be necessary to go higher or longer. However, when it forms part of a large capacity layout, the effect could be minimal.

OTHER LAYOUTS

- If the site is steeply sloping, a stepped SLD-type layout could be used to some advantage.
- For large-capacity layouts, a FSDR 1 layout is equally as efficient, but the rapid flow route is longer.

Figure VCM 3 Two-way-flow with a single end ramp

Intermediate and top deck levels

Scale 1: 600

Exit/entry level

Scale 1: 600

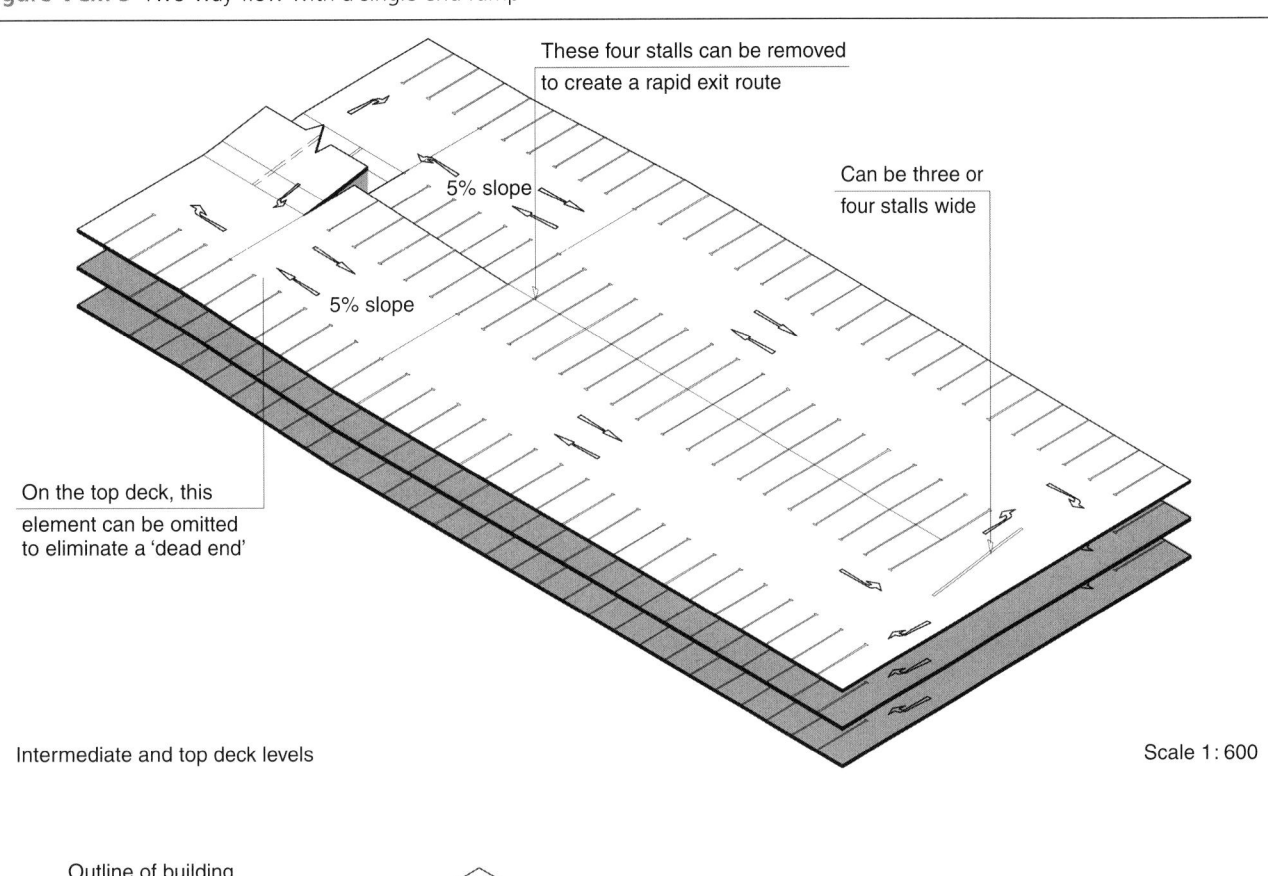

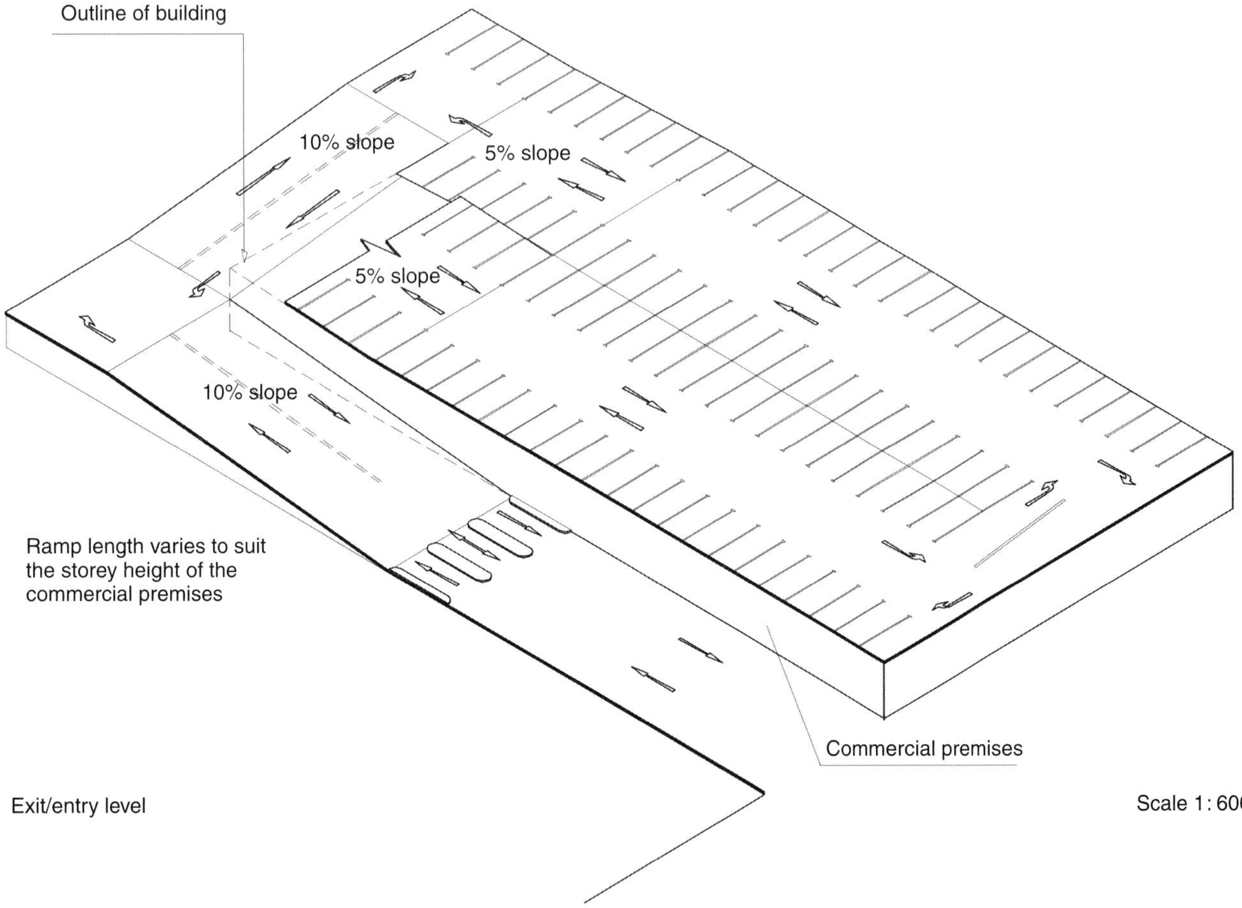

VCM 3 (Figure VCM 3)

ADVANTAGES

- As shown, it has been located above commercial premises where the mainly flat deck can be used to advantage. If located at ground level, it can also be accessed from the end of the flat aisles.
- Only one VCM module, suitably modified, is required.
- It is pedestrian-accessible throughout each deck without encountering slopes in excess of 5%.
- The storey height above the commercial premises can be varied by the length of the external sloping ramp elements.

DISADVANTAGE

If a 'dead end' is to be avoided, the top sloping element should be omitted.

COMMENTS

- At any time a rapid exit route can be introduced if it is found to be required.
- The vertical circulation route can be formed without any extra stall spaces being required, provided that there is no need for a rapid-flow route in either direction.
- Twenty-four stalls occur on the 5% sloping deck elements; all of the other stalls are located on a flat deck.
- Assuming that the ground conditions are flat and horizontal, the entry/exit position on the side of this layout is the most practical location. It has been shown to eliminate the problems associated with left-turning onto a control kiosk, although other fiscal methods may also eliminate that problem. This access location can also be used for a VCM 1 layout, provided that the flat deck area is big enough.
- Dependent on the overall building length, there can be a significant area of flat deck in the middle section and although shown as a two-bin layout, the modules can be located at any convenient position in a much larger deck area.
- It is not really suitable for small- to medium-capacity layouts where the VCM 1 layout with its flat ends is more flexible. It does, however, come into its own in large-capacity decks where the modules can be separated and relocated to other positions.
- Only eight stall spaces are required to complete the vertical circulation, regardless of the capacity of the deck.
- With the short sloping inflow aisle, angled parking has not been found to be necessary.

STATIC EFFICIENCY

- It is similar to VCM 1, but if 'dead ends' are to be avoided at ground level and the top deck, it might be necessary to go higher or longer. However, when it forms part of a large-capacity layout, the effect could be minimal.

OTHER LAYOUTS

- If the site is steeply sloping a stepped SLD-type layout could be used to some advantage.
- There are no other internal vertical circulation systems that are as efficient for large-capacity layouts.

Circulation layouts

Figure FSDR 1 One-way flow with internal ramp

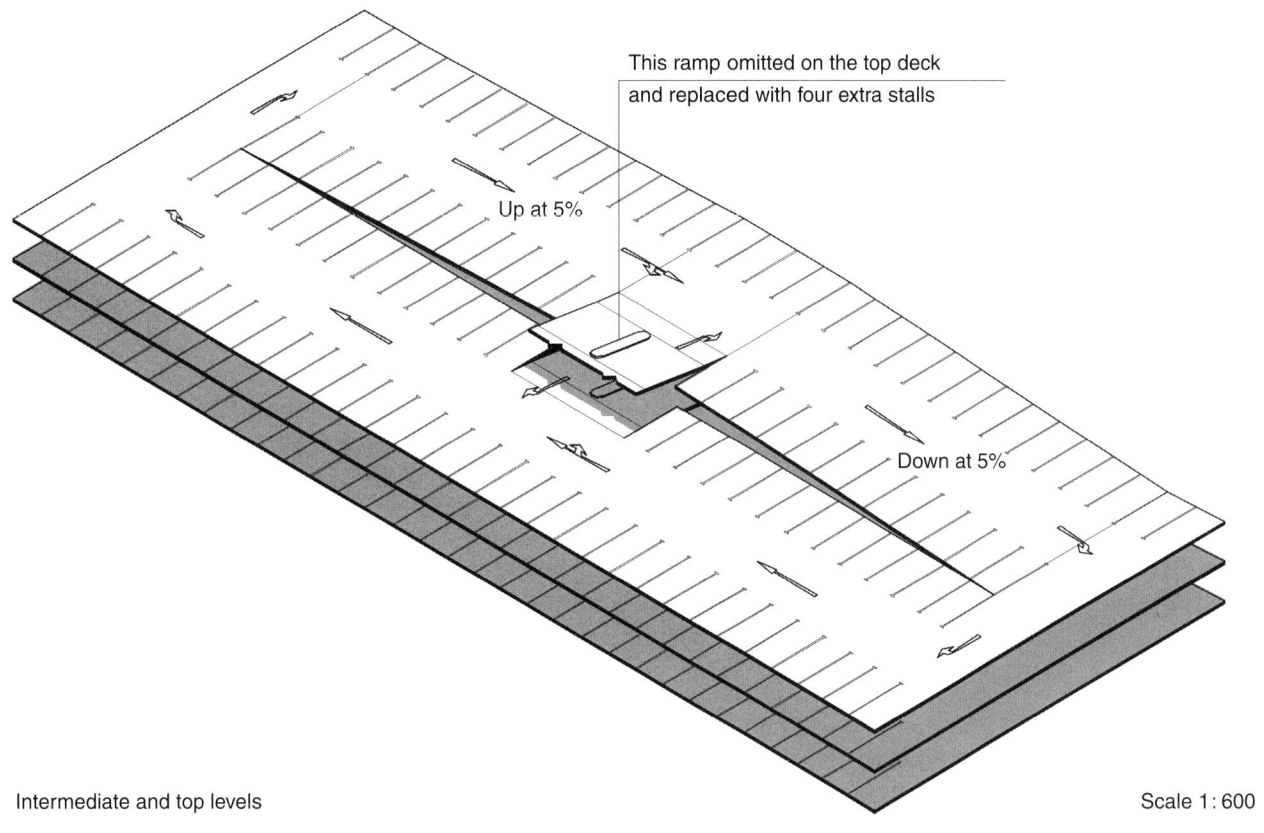

Intermediate and top levels

Scale 1:600

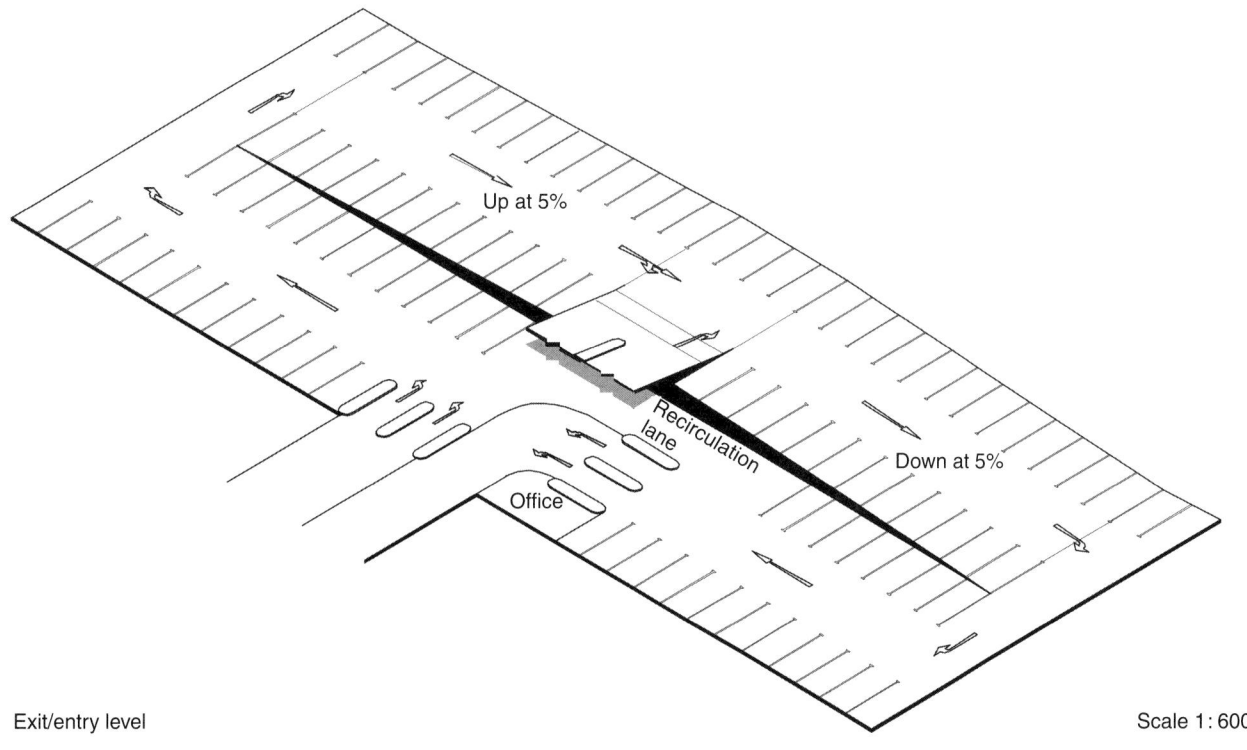

Exit/entry level

Scale 1:600

FSDR 1 (Figure FSDR 1)

ADVANTAGES

- It has a simple circulation and re-circulation capability.
- Storey heights can be varied by the length of the sloping deck elements.
- All turns are made in the same direction.

DISADVANTAGES

- It cannot be used in layouts less than 76.8 m (32 stall-widths) in length.
- At its minimum length, only 50% of the stalls can be searched on the inflow route.
- A minimum of 56 stalls per storey height need to be passed on the exit route.
- At its minimum length, pedestrian access between adjacent decks is limited to each aisle end.

COMMENTS

- The entry/exit location should be on the flat-deck side if economical construction considerations are to prevail. Entering from the other side requires altering the flow direction (see SD 4), and unless the sloping deck elements start from 1.5 m below ground level, the flat-deck element will not benefit from bearing on the ground.
- The sloping-deck elements are on the same aisle. However, the twelve-stall-width length of the sloping section of the inflow aisle renders it advisable to consider angled parking to prevent drivers from turning against the one-way traffic-flow when exiting the stalls (see SD 3 and 4).
- As shown, this layout with the cross-ways at the aisle ends only requires eight extra stall spaces to complete the vertical circulation. With longer layouts, 12 or 16 may be required, but even so, this is quite acceptable.
- It is not really suitable for smaller capacity layouts, but when used for large-capacity facilities, the reduction in cross-deck pedestrian access may not be so significant.

STATIC EFFICIENCY

- As shown, there are 112 stalls. The area per car space is 21.4 m^2 and can be deemed 'good'.

OTHER LAYOUTS

- A VCM 1 layout of the same length as shown here is also of similar dynamic efficiency, but with over 50% of its deck area laid flat across both bins it is more 'user friendly' and economical to construct at ground level.
- SD 3 and 4 layouts are equally as efficient and have better cross-deck access for pedestrians by virtue of the central cross-way.

Circulation layouts

Figure WPD 1a Warped Parking Decks

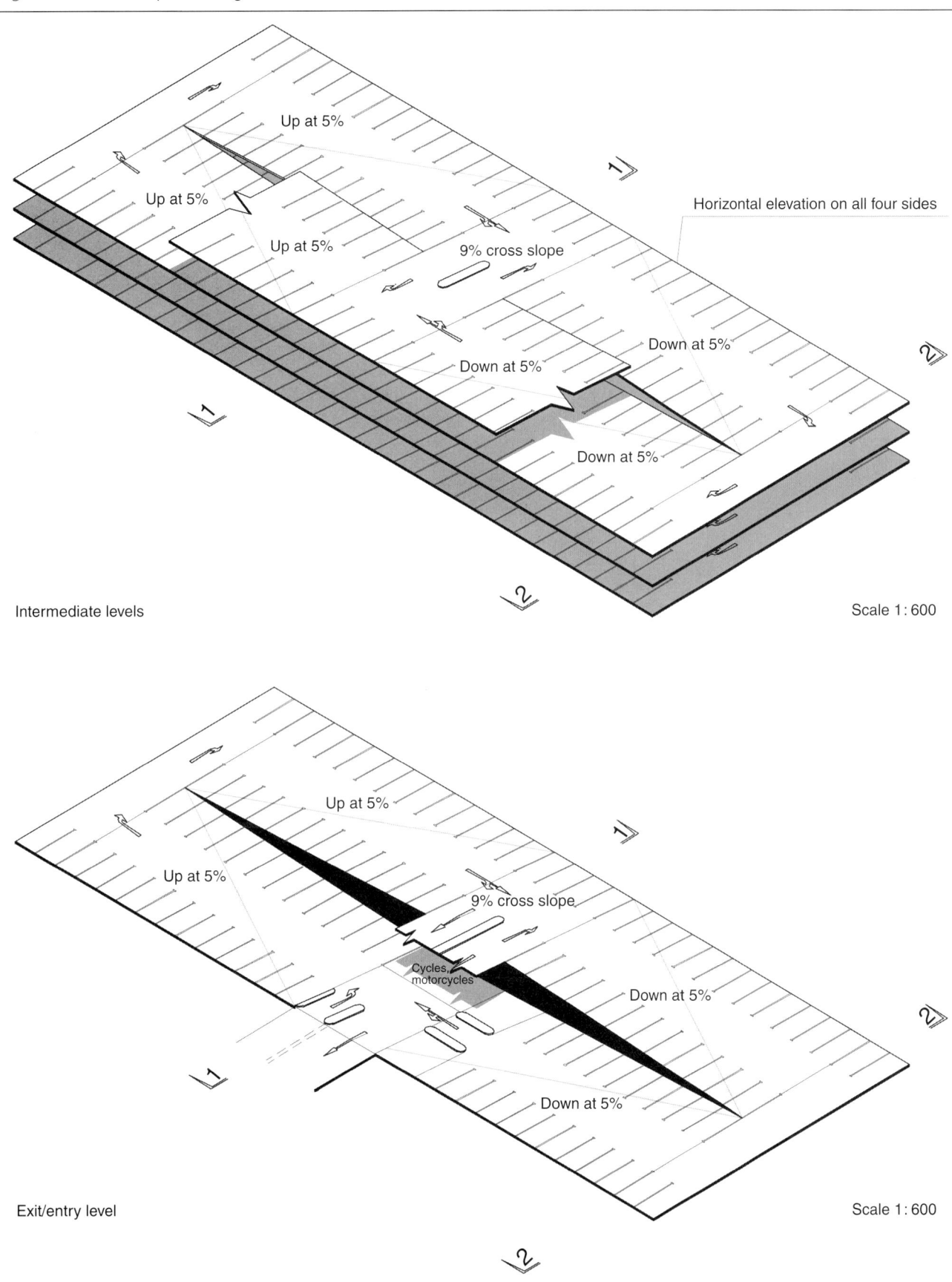

Figure WPD 1b Warped Parking Decks

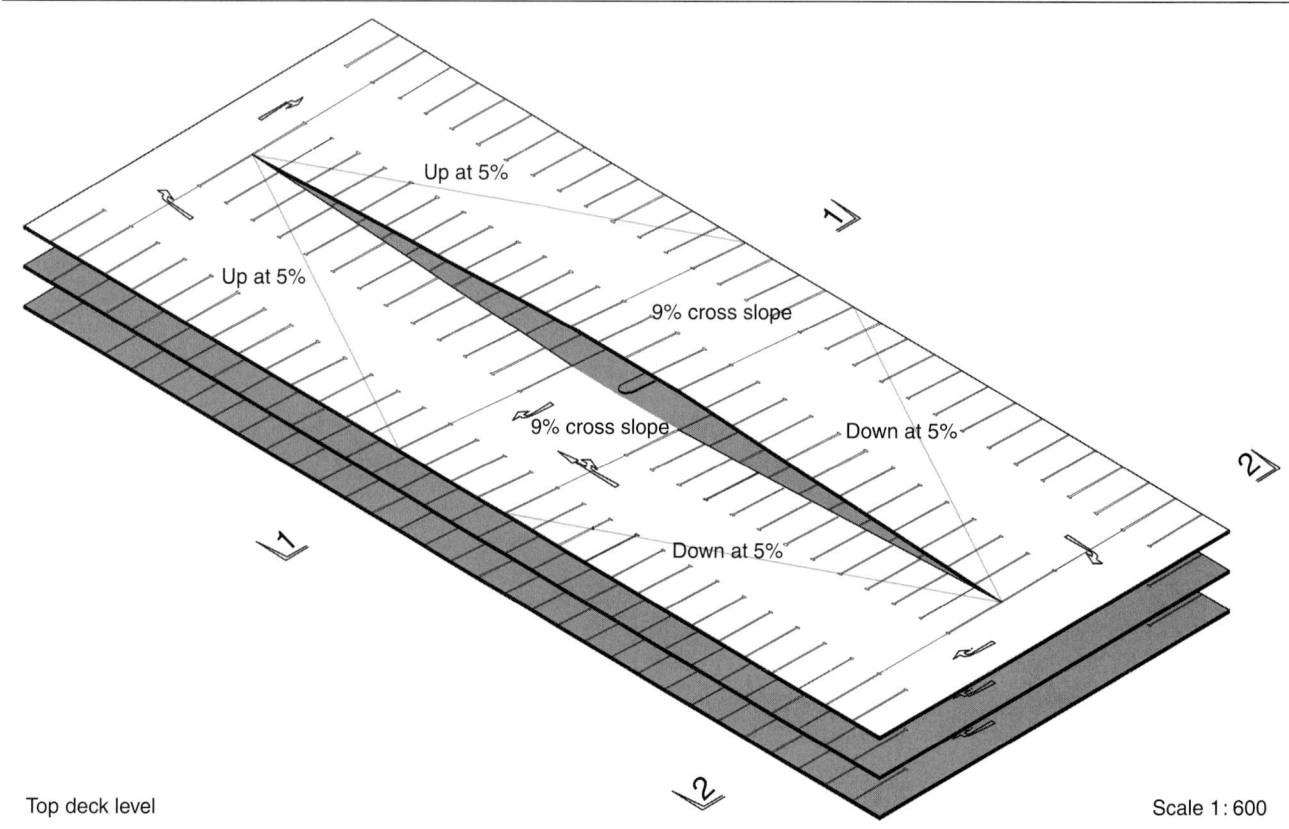

Top deck level

Scale 1:600

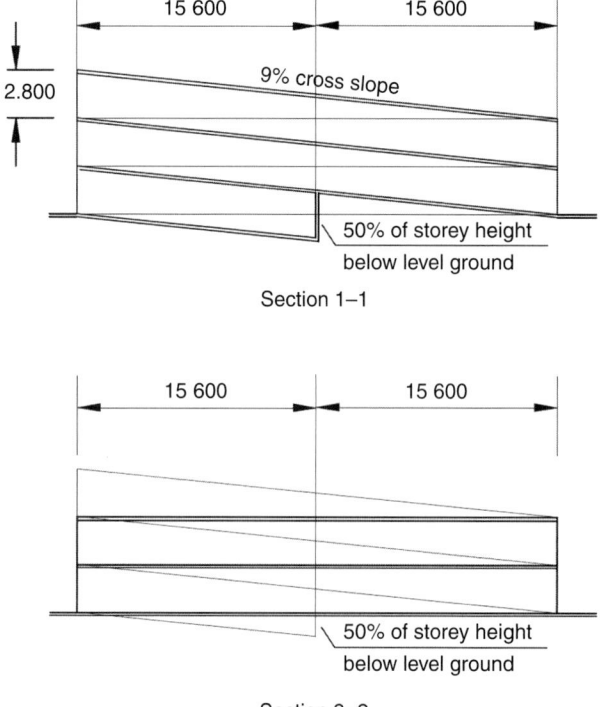

WPD 1 (Figures WPD 1a and 1b)

ADVANTAGES
- The deck elevations are horizontal on all four sides.
- Pedestrian access between adjacent bins occurs at each end of the traffic aisles.
- It incorporates a simple re-circulation capability.

DISADVANTAGES
- The centrally located vehicle ramp in the middle of the layout is too steep to allow pedestrian use.
- Maximum allowable dimensions for pedestrian slopes are also exceeded on the parking decks (see BS 8300:8.2 (BSI, 2010)).
- A minimum length of 74.4 m (31 stall-widths) is required when using 5% sideways parking slopes, but it is to be appreciated that the slope in the other direction can be in excess of 9%.
- 50% of the stalls are located on the extended outflow route.
- Fully-laden shopping trolleys are difficult to control on parking decks that incorporate falls in excess of 5%, especially when the aisle slopes are diagonal.

COMMENTS
- Sixteen stall spaces per deck are used to complete the circulation route (14 if a combined ramp is adopted).
- The decks are horizontal at each end of the building and warp to a maximum at the central ramp locations.
- This layout was much used 30 to 40 years ago. Examples of this type of car park have been constructed throughout the UK, and some of them are still in operation.
- Recent legislation has reduced the maximum slope allowed for pedestrians on ramps.

STATIC EFFICIENCY
- As drawn, the number of stalls is 112, and the static efficiency, at 21.43 m^2 per car space, can be deemed 'Good'.

OTHER LAYOUTS
- Layouts in the SD series share similar circulation features, especially SD 3, which incorporates an identical circulation layout, and they are more user-friendly in having a flat central access-way for pedestrians. They cannot, however, be constructed with the deck sides horizontal.
- A VCM 1 layout is more user-friendly but without warping the decks it cannot be constructed with the building sides horizontal. However, the maximum deck slope in either direction does not exceed 5%.

7.9. Flat parking decks with storey height internal ramps (FIR)

In layouts three or more bins wide, internal cross-ramps can be introduced which climb through a complete storey height. In the UK, this has been used, in the main, to create a horizontal elevation on all four sides of a multi-bin system, without resorting to sloping or stepped parking decks.

Two basic types occur, those with ramps running across the bins and those where the ramps run parallel to the traffic aisles. The slope of the internal ramp is recommended to be a maximum of 10%. Designers of existing buildings with cross-ramps have, in general, increased the ramp slopes slightly to fit the parking dimensions with no known complaints from motorists.

When used across the decks the dimension available for a ramp is usually 25.2 m. In the other direction, the length is unrestricted and can be 'tailored' to suit any storey height. The cross-ramp location results in an interruption to the free flow of traffic along the central bin, resulting in a loss of stalls if 'dead ends' are to be avoided.

Dynamic and static efficiency is not a good feature of this layout type (see Chapter 4). The layouts featured show that drivers need to pass through some traffic aisles and access-ways more than once, in order to search all of the stalls on any particular level.

Circulation layouts

Figure FIR 1 One-way flow with full height internal ramps

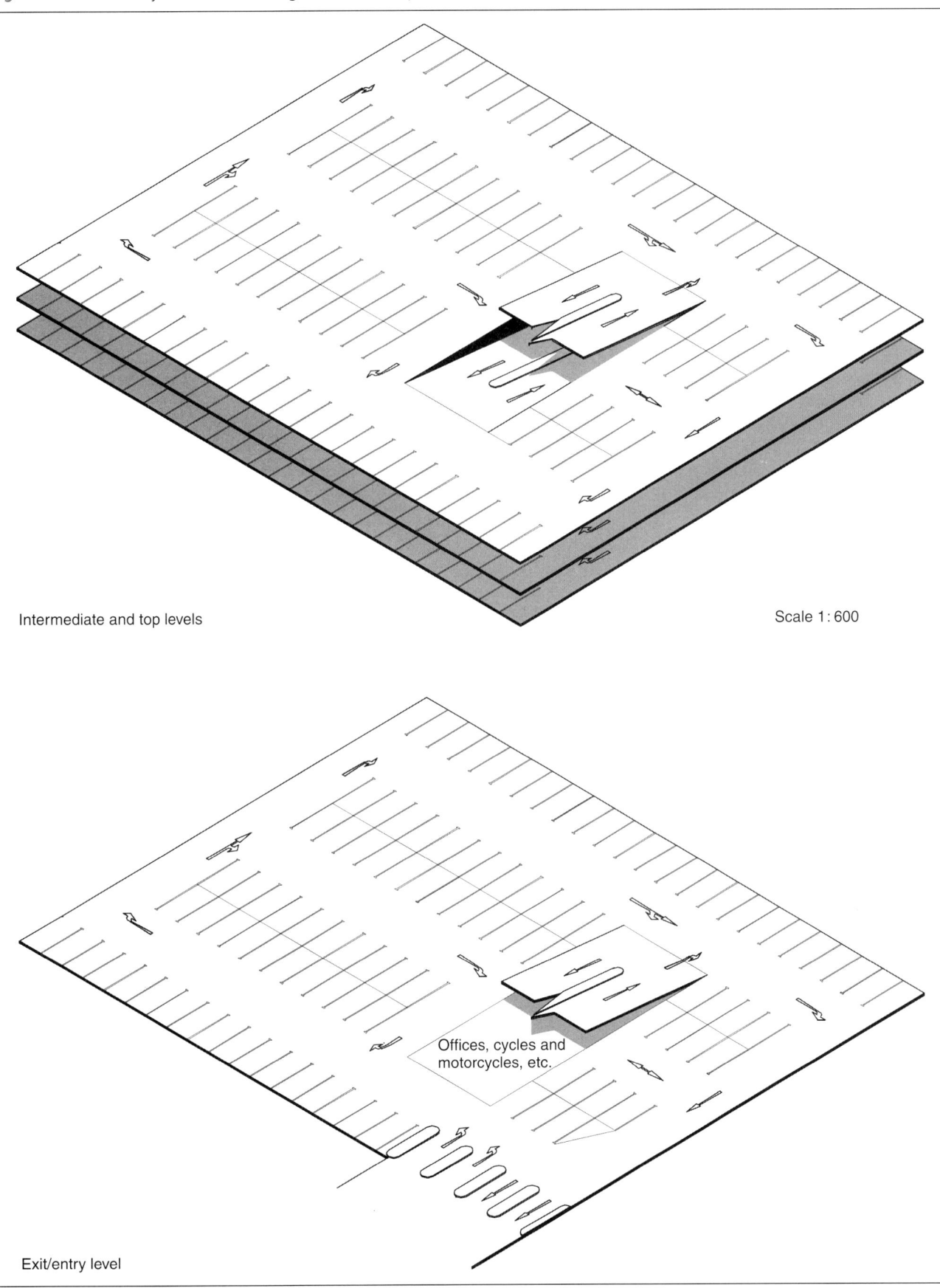

Intermediate and top levels

Scale 1:600

Exit/entry level

Offices, cycles and motorcycles, etc.

FIR 1 (Figure FIR 1)

ADVANTAGES

- There are horizontal elevations to all four sides.
- All turns are in the same direction with no single turn greater than 90°.
- It incorporates a simple re-circulation capability.
- There is flat access for pedestrians between adjacent bins.
- The outflow route is reasonably rapid.

DISADVANTAGES

- The internal ramps climb through a full storey height. Their length, between aisles, is 25.2 m; the maximum recommended slope is 10%.
- The ramp crosses the central bin, creating dead ends.
- Thirty-two stall spaces per deck are required to complete the circulation route (36 spaces if all dead ends are to be avoided).
- The static and dynamic efficiency is not good, with many stalls having to be passed twice as each deck level is searched.

COMMENTS

- If the maximum allowable ramp slope of 10% is not exceeded, the storey height should not be greater than 2.56 m. For increased storey heights, it will be necessary to extend the central bin dimension or create gaps between adjacent bins.
- If all of the stalls are to be searched on each level and motorists drive up to the next parking level, they must drive more than twice the distance around the length of the perimeter aisles and access-ways. Circulation efficiency is low and can be a major cause of traffic congestion at busy times.
- An alternative is to drive up to the top parking level on the inflow route, transfer to the outflow route, and continue searching on the way back down. Even so, on the inflow route one of the aisles has to be driven over twice.
- The introduction of a variable message sign system can improve circulation efficiency, but the extended length access-ways and ramps and the need, still, to drive through some aisles more than once, do not make this layout as efficient as most of the others.
- The high circulation requirement of up to 40 stall spaces per deck is a poor feature.
- The minimum aisle length is 36 m (15 stalls), but at this length more than 30% of the stall spaces will be required to complete the circulation pattern.
- The low circulation-efficiency renders it unsuitable for any parking category where intensive activity is anticipated.

STATIC EFFICIENCY

- As drawn, the number of stalls is 112 and the static efficiency, at 24.07 m^2 per car space, can be deemed 'average'.

OTHER LAYOUTS

- A VCM 1 layout, three bins wide, has superior static and dynamic efficiencies without dead ends, and is more user-friendly.
- If there were sufficient space on the site, any of the external ramp systems would also produce superior layouts.

Circulation layouts

Figure FIR 2 One-way flow with express ramps

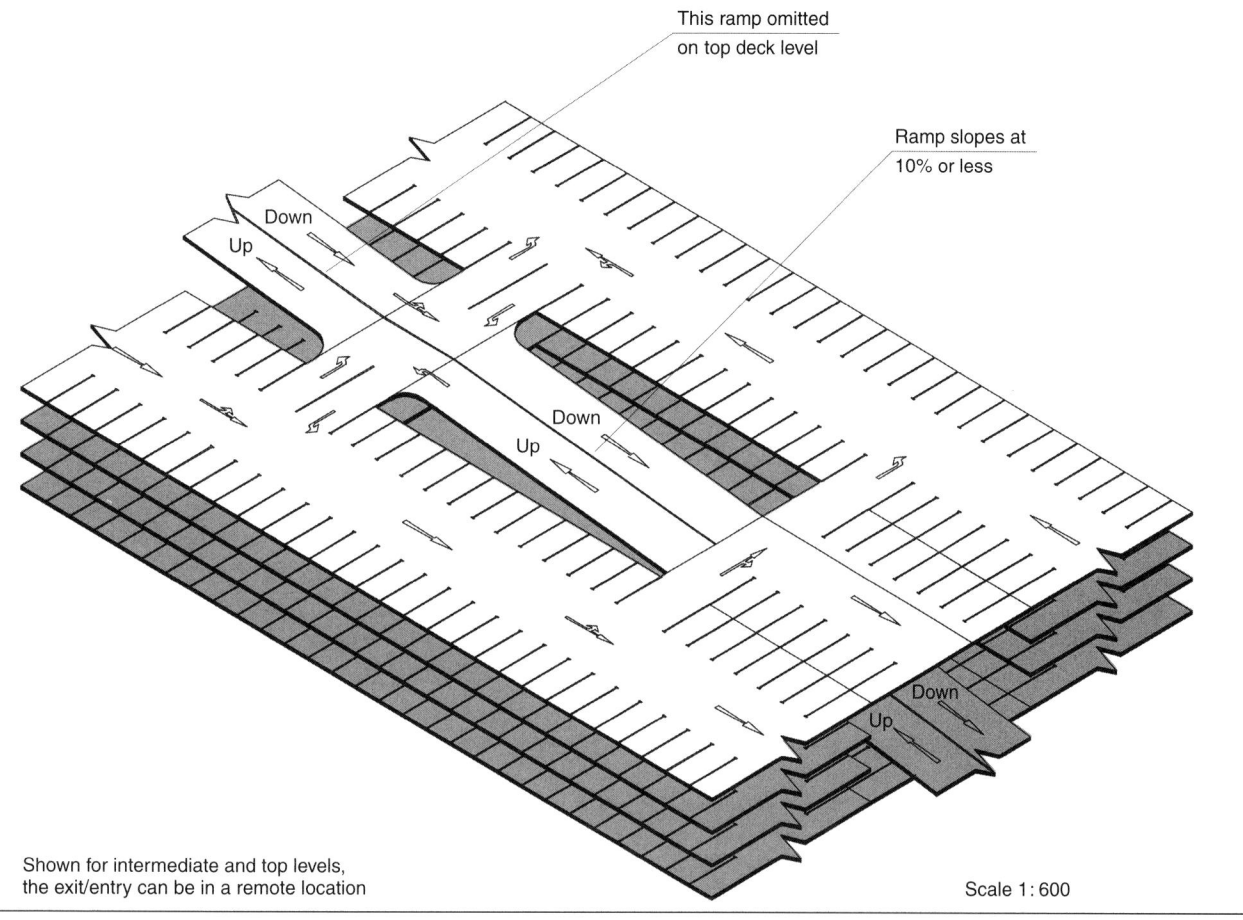

Shown for intermediate and top levels, the exit/entry can be in a remote location

Scale 1:600

Figure FIR 3 One-way flow with circular ramps

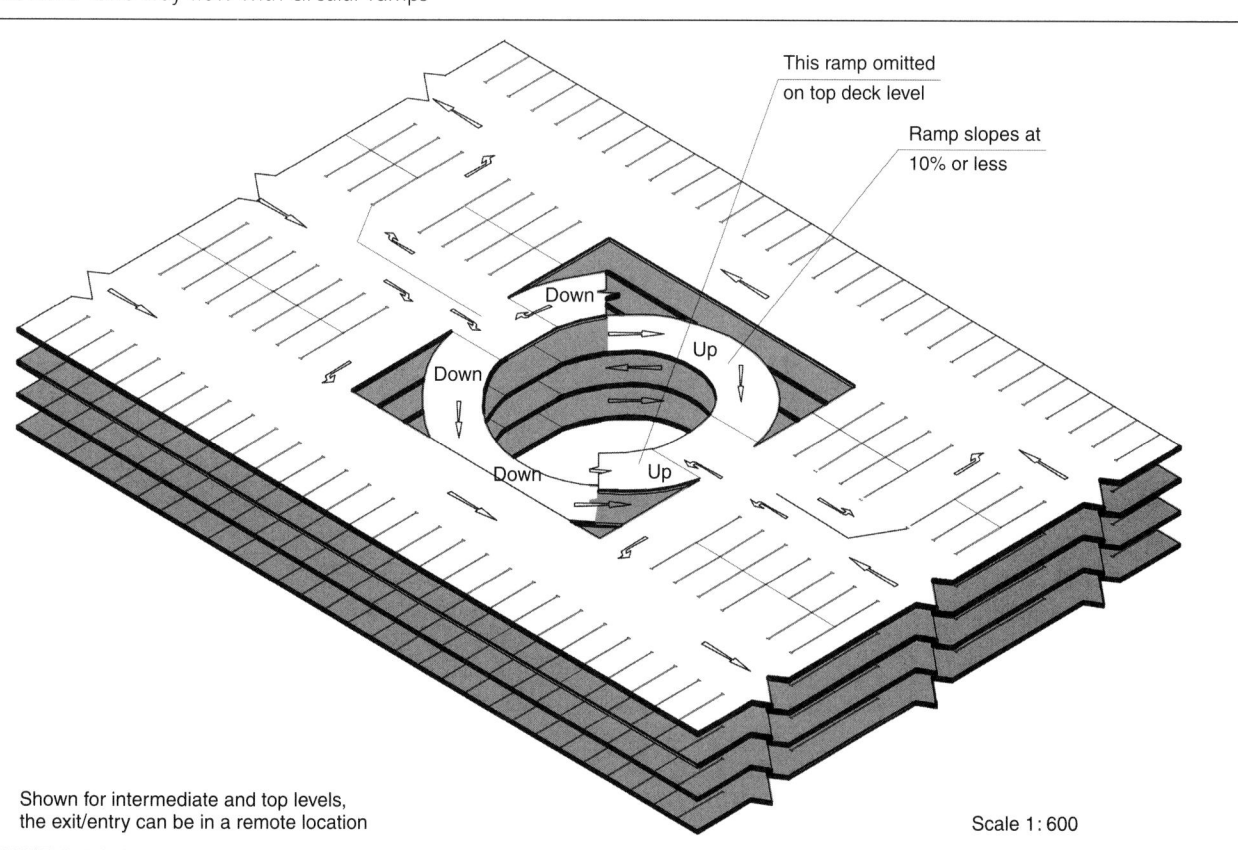

Shown for intermediate and top levels, the exit/entry can be in a remote location

Scale 1:600

FIR 2 and 3 (Figures FIR 2 and 3)

ADVANTAGES
- It incorporates flat decks.
- It provides a simple re-circulation capability.
- There is flat access for pedestrians.
- Inflow and outflow is rapid.
- Congested floors can be bypassed with the aid of traffic marshals or variable message signs.
- There are no dead ends.

DISADVANTAGE
It is only suitable for larger facilities since the ramps take up a large part of the floor area.

COMMENT
These layouts are particularly suitable for large tourist-venue-type car parks where it is necessary to deal with high flows of traffic entering and leaving over a very short period of time.

STATIC EFFICIENCY
- Static efficiency will be dependent on the ratio of the ramp area to the total floor area but it is likely to be in excess of the 28 m^2 per car space described for FIR 3.

Figure FIR 4 One-way flow with edge ramps

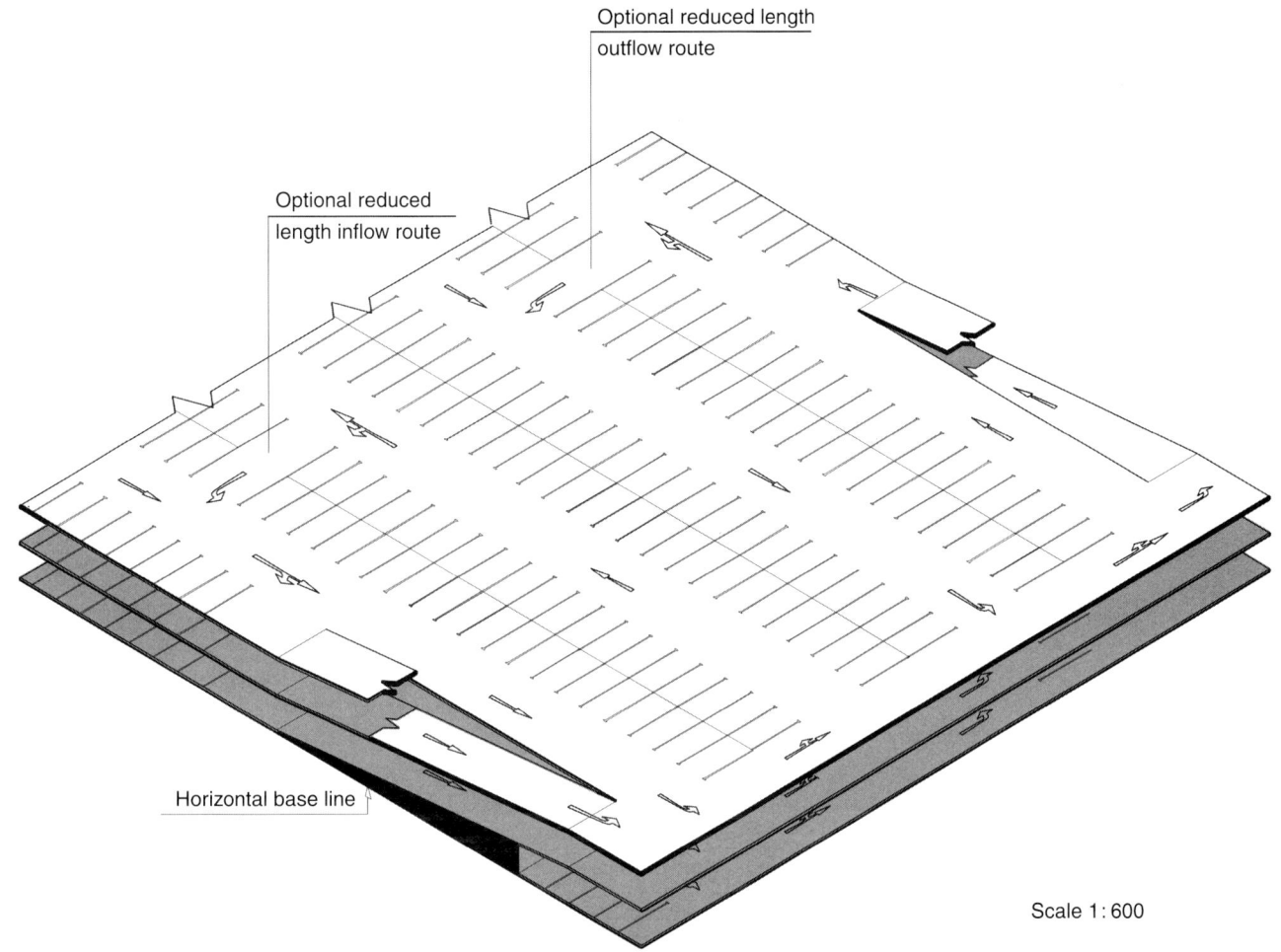

Scale 1:600

FIR 4 (Figure FIR 4)

ADVANTAGES
- There are horizontal elevations on all four sides.
- It has a simple re-circulation capability.
- Flat access is provided for pedestrians between adjacent bins.
- The outflow route is reasonably rapid.
- The ramps are located at right angles to the direction for a type FIR 1 layout. This enables a 10% sloping ramp to be introduced for any particular storey height.
- Dead ends are eliminated.

DISADVANTAGES
- The high circulation requirement of 44 stall spaces per deck is a poor feature and can only really be justified by being incorporated within a much larger deck layout.

COMMENTS
- The layout is uncomplicated, and the traffic pattern is simple to understand and is more efficient than that for an FIR 1 layout.
- The higher circulation requirement should be weighed against the dynamic advantages when compared with an FIR 1 layout.

STATIC EFFICIENCY
- Dependent on the capacity of the deck within which it is incorporated, the overall static efficiency will vary. However, if constructed as a small independent layout, static efficiency will be in the order of 28 m^2 per car space. This can only be described as 'poor'.

OTHER LAYOUTS
- A VCM 1 layout, three bins wide, has superior static and dynamic qualities.
- If there were sufficient space on the site, any of the HER and ER series could also produce superior layouts.

7.10. Minimum dimension layouts (MD)

The smallest practical size for any parking layout is dictated by the recommended minimum turning circles for the SDV. It does not necessarily have to be over the full length of the building but at the ends, at least, in order to achieve a turning dimension. In continental Europe, 'ring spanner'-shaped layouts on several levels (MD 1) occur under the main shopping streets of some towns, for example, Rheims.

They are useful for very long and narrow sites both above and below ground and as they increase in length so their static efficiency improves, but there are no facilities of this type known to occur in the UK at the present time.

A two-bin, SLD 3-type car park can be constructed down to a plan-size of 31.2 m × 24.0 m, dimensions that can only be matched by an SLD 4 and a VCM 3 layout. Stalls 2.3 m wide reduce this dimension even further and if the parking need is great, the site is small and the client is amenable, such a reduction in parking standards may well be acceptable.

A long MD 1-type layout can have a greater static efficiency than any other circulation design (<20 m^2 per space) followed closely by an MD 2 layout, when discounting ramps. For the other layouts in the MD series, increases in the building length actually reduce static efficiency due to the internal bins having the same deck area as those located at each end, but containing fewer stalls.

The static efficiency of the MD 3 to MD 5 types compares, unfavourably, with most other layout types (28 to 30 m^2 per car space). However, as there are no other types that can be used in their place, any comparison is academic.

Figure MD 1 One-way flow between end ramps

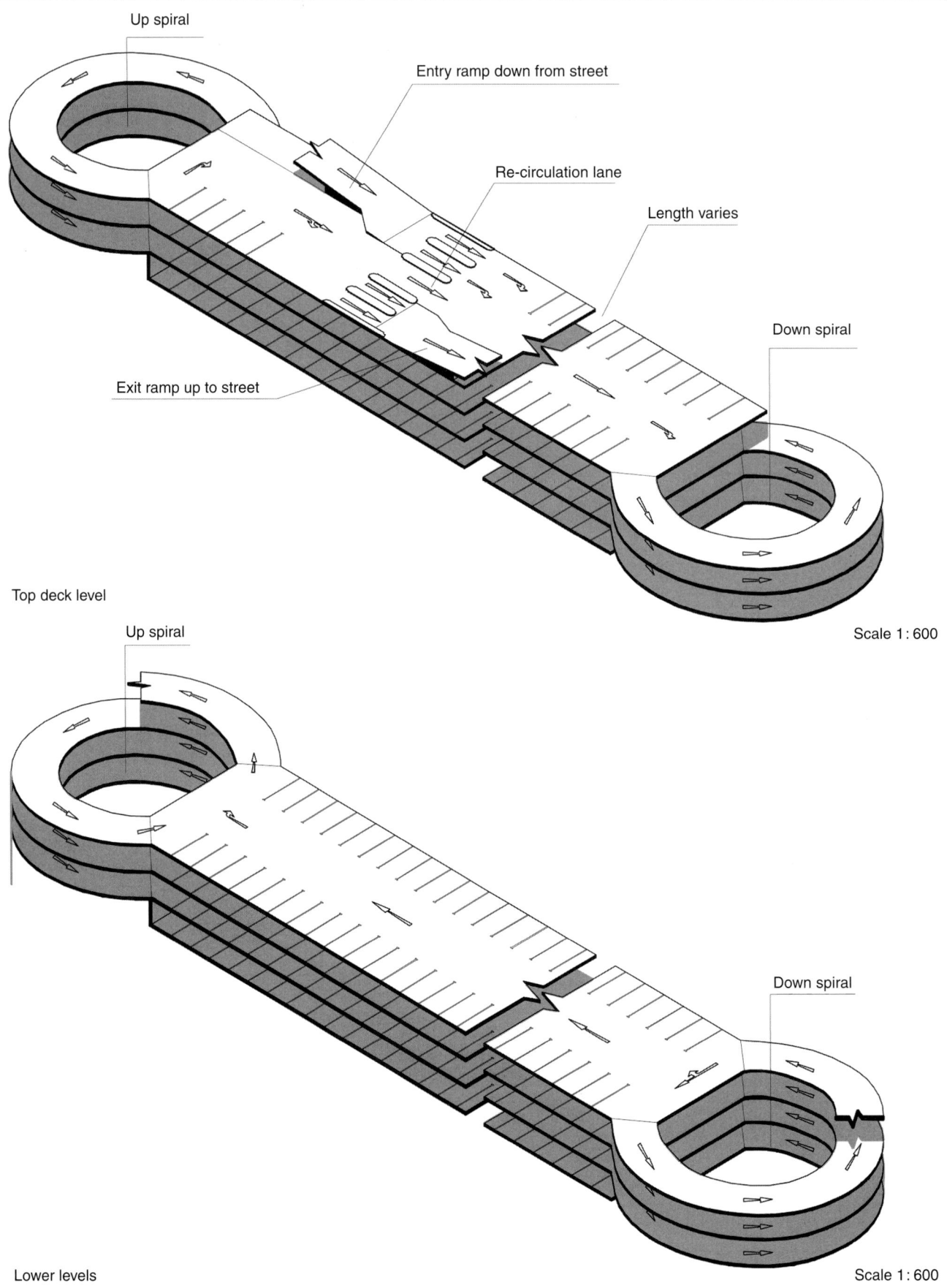

Circulation layouts

Figures MD 1a and 1b End ramp variations

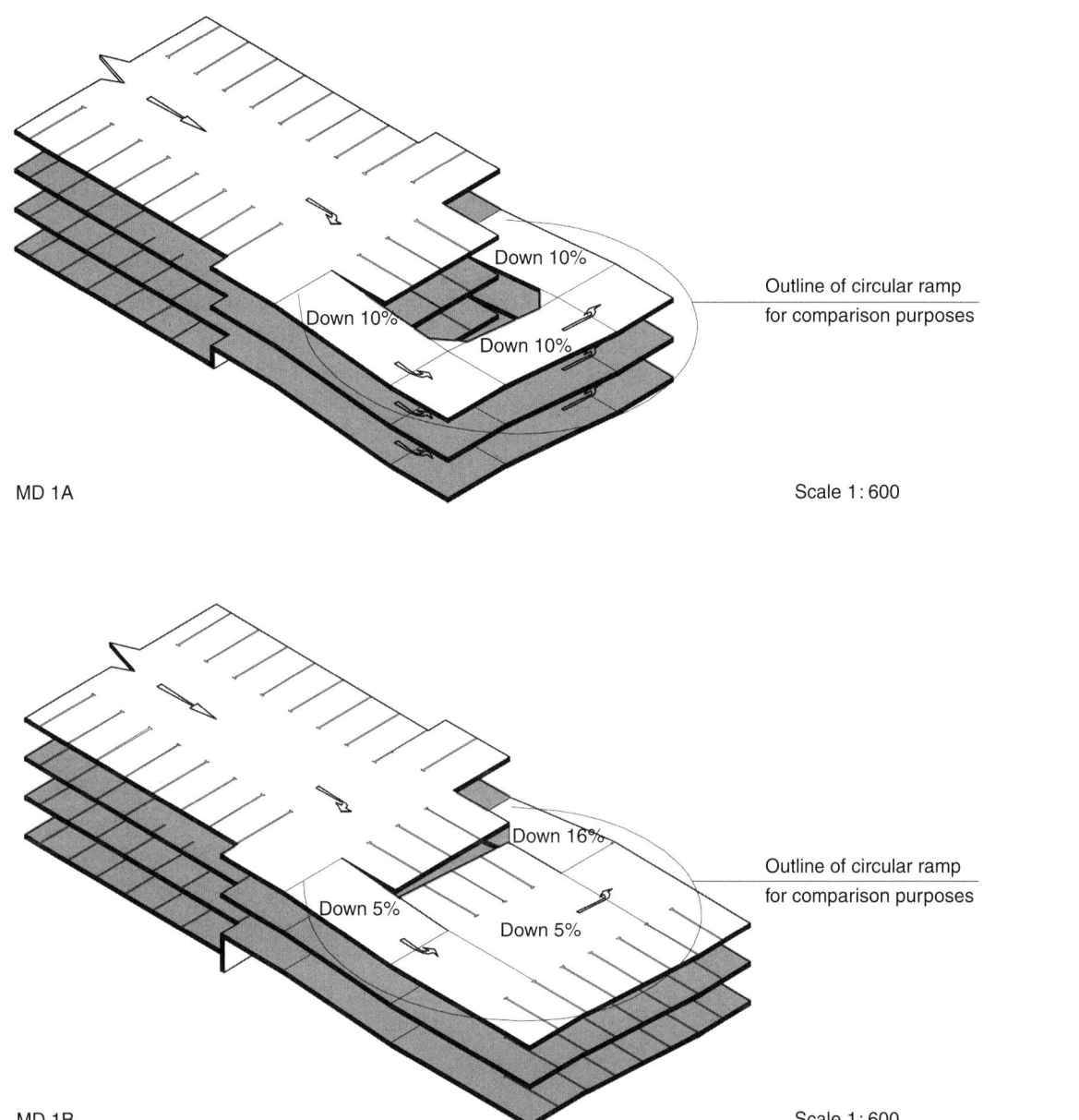

MD 1A Scale 1:600

MD 1B Scale 1:600

MD 1 (Figures MD 1, 1a and 1b)
ADVANTAGES
- It can be accommodated on minimal width sites (19 m at the ends and 16 m in the central section), either above or below ground level.
- Provided that the parking status of the decks can be shown (VMS), full or congested levels can be bypassed enabling drivers to proceed directly to a level where stalls are readily available.
- With the circular or MD 1A ends, the longer the aisle length becomes, the more statically efficient it becomes, but when MD 1B ends are used, the efficiency remains relatively constant regardless of the aisle length.

DISADVANTAGE
The traffic circulation and re-circulation route is vertical rather than horizontal.

COMMENTS
- As shown, the circulation is all one-way, in the same direction, and requires the use of a VMS if it is to work efficiently. There is, however, an alternative. Regarding each pair of decks as a vertical rather than a horizontal circulation system, with alternate decks flowing in opposite directions, then the decks can be searched in pairs. It does mean that, if unsuccessful, drivers must return to the deck two levels above (for an underground facility) and come back down three levels to start the next search: therefore it is not recommended.
- The only known layout utilising this layout is in Rheims and was constructed under a main shopping street. There are no known facilities of this type operating elsewhere in the UK or mainland Europe.

STATIC EFFICIENCY
As shown, with circular ends, the area per car space for the MD 1 and 1A is 28 m^2 and when the aisle length is extended to contain 100 stalls, it improves to 24 m^2. Similar-capacity MD 1B figures are 24.0 m^2 and 22.8 m^2 respectively.

OTHER LAYOUTS
- If a building-width of 19.2 m can be accommodated, then a MD 5 layout would be dynamically superior, but it becomes less statically efficient as the aisle length increases.
- For layouts with aisle lengths of more than 50 m in length, a MD 1B layout is more efficient, dynamically, but not as efficient, statically.

Circulation layouts

Figure MD 2 Two-way flow with one end ramp

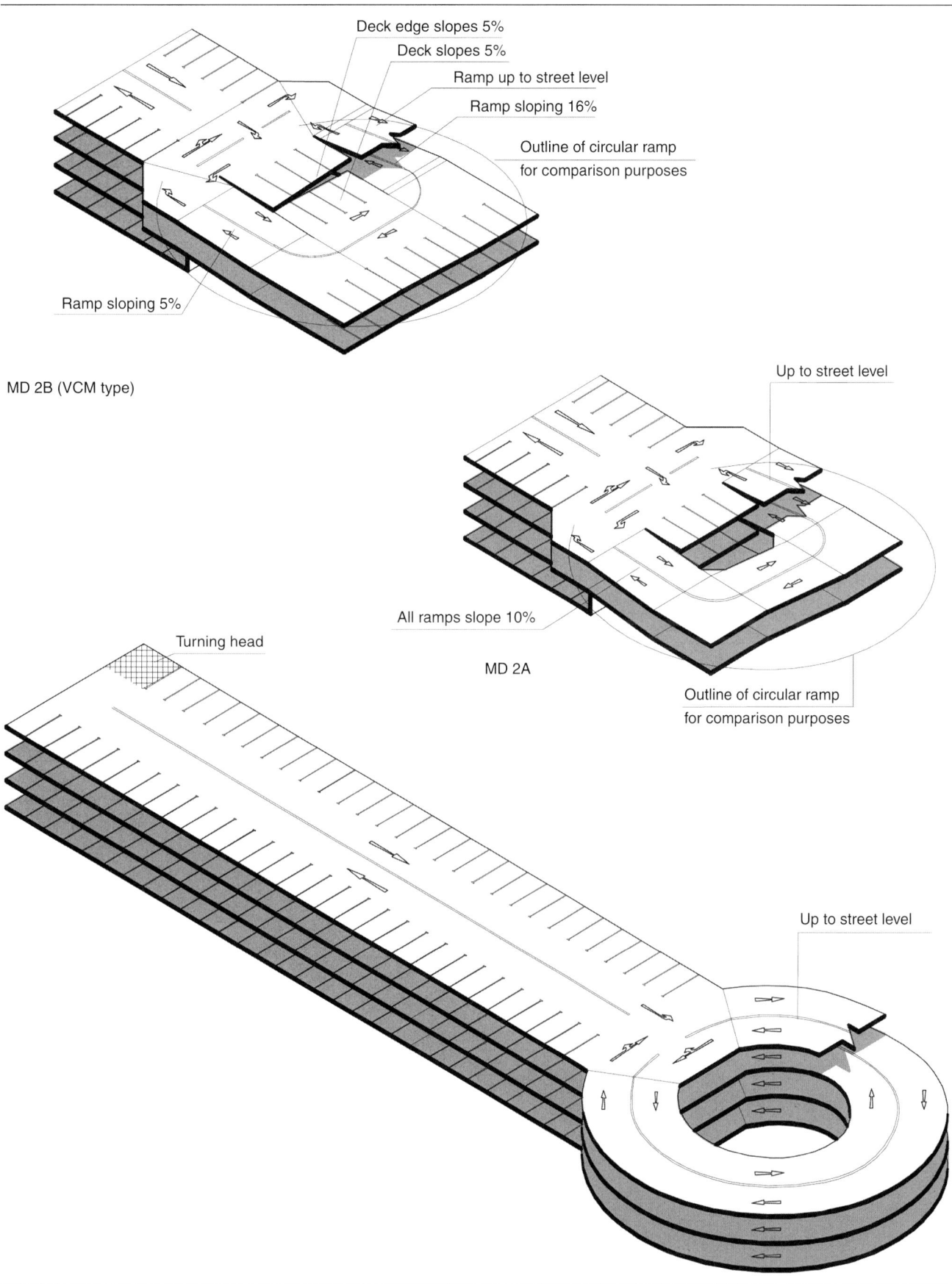

MD 2 (Figure MD 2)

ADVANTAGES

- It can be accommodated on narrow width sites (27 m at the ends and 17 m in the central section), either above or below ground level.
- Each deck can be searched in succession, up or down.
- With the circular- or 2A-type ends, the longer the aisle length, the more statically efficient it becomes, but when 2B ends are used, the efficiency remains relatively constant regardless of the aisle length.
- The ability to flow in both directions on the same deck renders this more efficient, dynamically, than a MD 1 layout.

DISADVANTAGES

- The entry to each parking deck involves a cross-over condition.
- The extra width required for the two-way flow aisle renders this slightly less efficient, statically, than a similar-length MD 1 layout.

COMMENTS

- The circular end is shown simply as a comparison with the MD 1 layout. The original proposal (under a town centre in Portugal), employed MD 2B-type ends that rendered it more efficient.
- Although the circulation pattern works without the use of a VMS, such a system is highly recommended if driver frustration is to be avoided. Without it, the longer the aisle, the greater will be the frustration.
- The 2B layout-end is 24 m wide, which is 3 m less than that for the circular option.

STATIC EFFICIENCY

As shown, with circular ends, the area per car space for the MD 2 and 2A is 27 m^2 and when the aisle length is extended to contain 100 stalls, it improves to 23.5 m^2. Similar capacity MD 1B figures are 24 m^2 and 23 m^2 respectively.

OTHER LAYOUTS

- If a building width of 19.2 m can be accommodated then a MD 5 layout would be dynamically superior, but it becomes less statically efficient as the aisle length increases.
- For layouts with aisle lengths of more than 50 m in length, a MD 1B layout is more efficient, statically, but not as efficient, dynamically.

Circulation layouts

Figure MD 3 SLD type. Ten stalls wide, with one end ramp

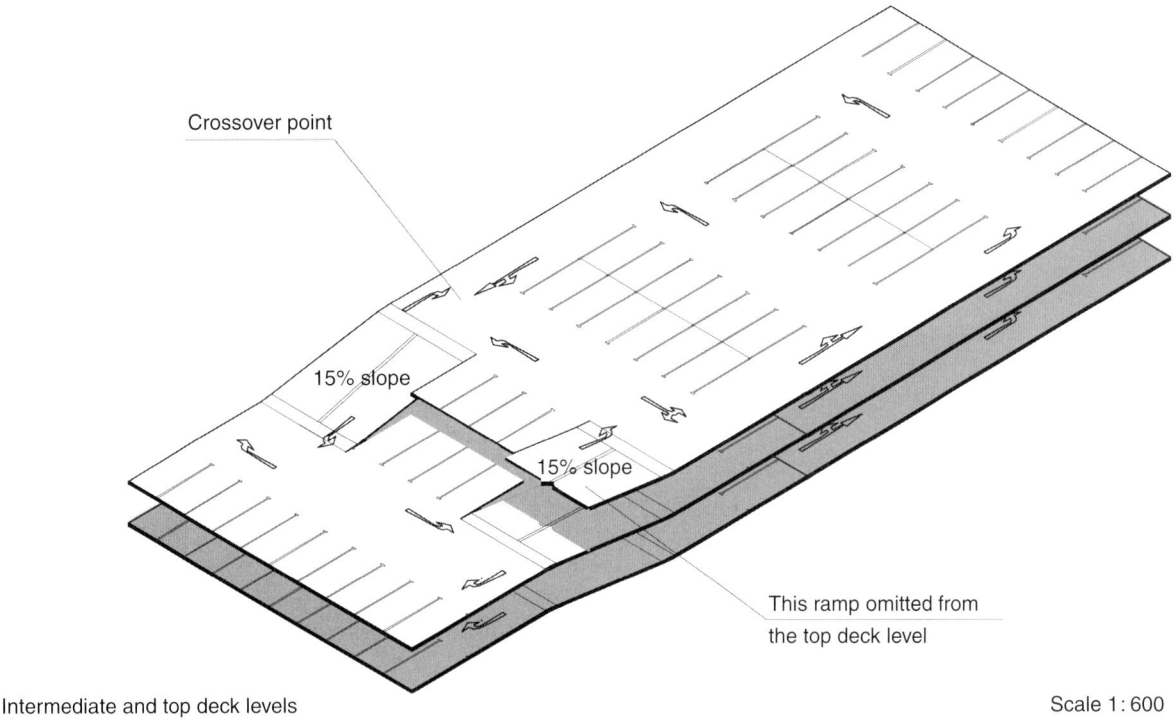

Intermediate and top deck levels

Scale 1:600

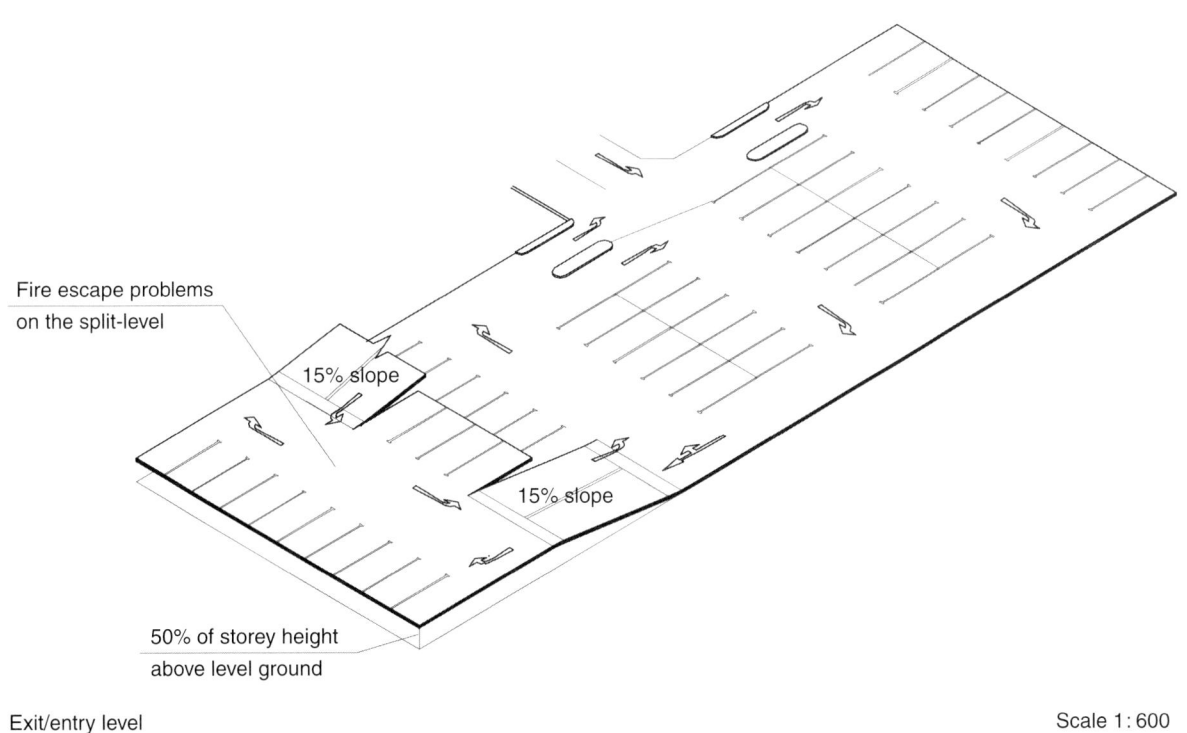

Exit/entry level

Scale 1:600

MD 3 (Figure MD 3)

ADVANTAGES

- It can be accommodated on sites down to 24 m in width, either above or below ground level.
- Each deck can be searched in succession, up or down.
- It has good re-circulation capabilities.

DISADVANTAGES

- Its static efficiency is not good and actually reduces as the building increases in length.
- The split-level end will also need to be provided with a fire escape.

COMMENTS

- The wide, two-way flow ramp makes it suitable for 'tidal' flow situations.
- The high proportion of cross-ways to stalls for the internal bins makes for a reduced static efficiency when compared with long MD 1 and 2 layouts. However, when the building length is less than 93.6 m, it can be similar and when only 48.8 m, it can be superior.

STATIC EFFICIENCY

Although the static efficiency is not good by normal standards, this layout can be constructed down to a length of 33.2 m at which dimension, comparison with other layouts is academic.

OTHER LAYOUTS

Apart from a MD 2B layout, there are no other layouts of similar size with the vertical circulation located at one end.

Circulation layouts

Figure MD 4 VCM type. Eight stalls wide

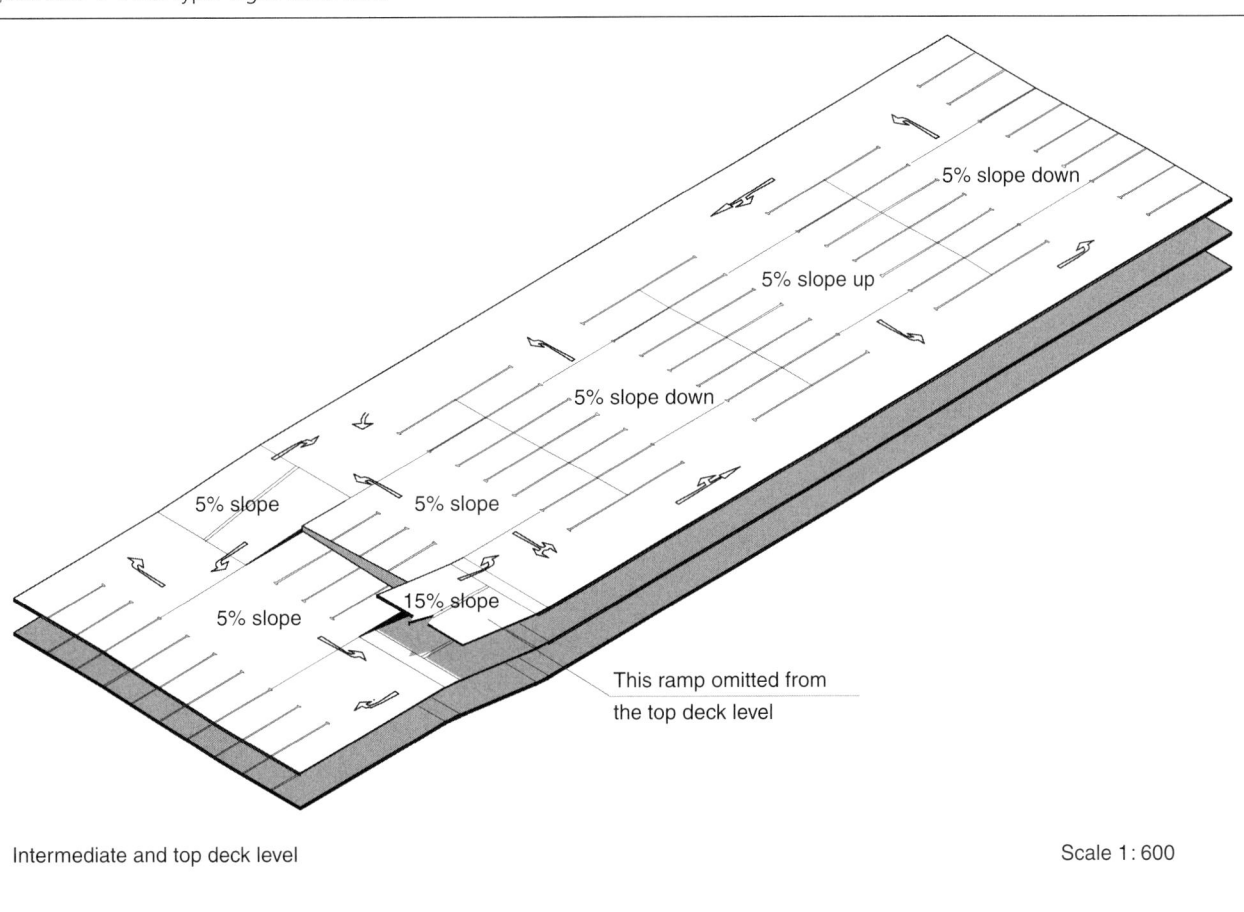

Intermediate and top deck level

Scale 1:600

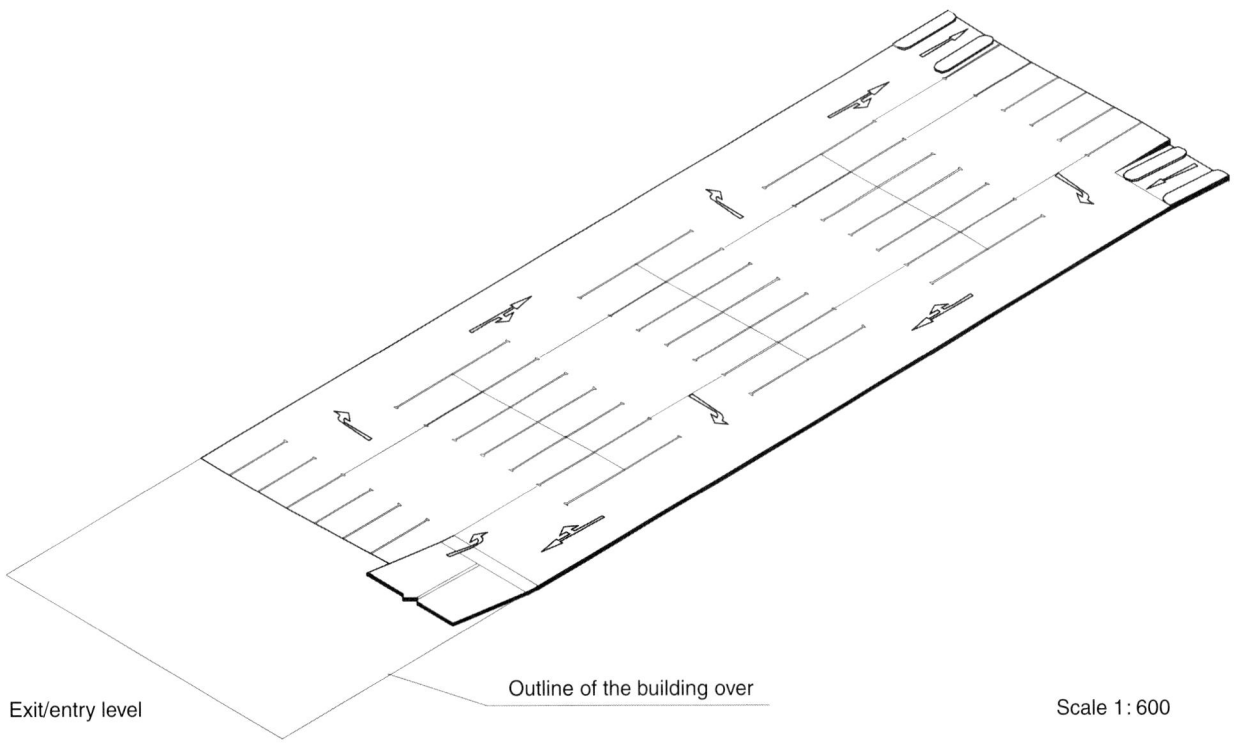

Exit/entry level

Scale 1:600

MD 4 (Figure MD 4)

ADVANTAGES

- It can be accommodated on sites down to 24 m in width, either above or below ground level.
- Each deck can be searched in succession, up or down.
- It has good re-circulation capabilities.
- A separate fire escape for the 'split level' end bin is not required.

DISADVANTAGES

Its static efficiency is not good and actually reduces as the building increases in length.

COMMENTS

- This is a VCM variation of the MD 3 layout that enables pedestrians to gain access to all parts of a deck without encountering a slope in excess of 5%.
- The wide, two-way flow ramp makes it suitable for 'tidal' flow situations.
- The high proportion of cross-ways to stalls for the internal bins makes for a reduced static efficiency when compared with long MD 1 and 2 layouts. However, when the building length is less than 93.6 m, it can be similar and when only 48.8 m, it can be superior.

STATIC EFFICIENCY

Although the static efficiency is not good by normal standards, this layout can be constructed down to a length of 33.2 m, at which dimension, comparison with other layouts is academic.

OTHER LAYOUTS

- Apart from a MD 2B or MD 3 layout, there are no other layouts of similar dimensions and with the vertical circulation located at one end.

Circulation layouts

Figure MD 5 VCM type. Eight-stalls-wide with two end ramps

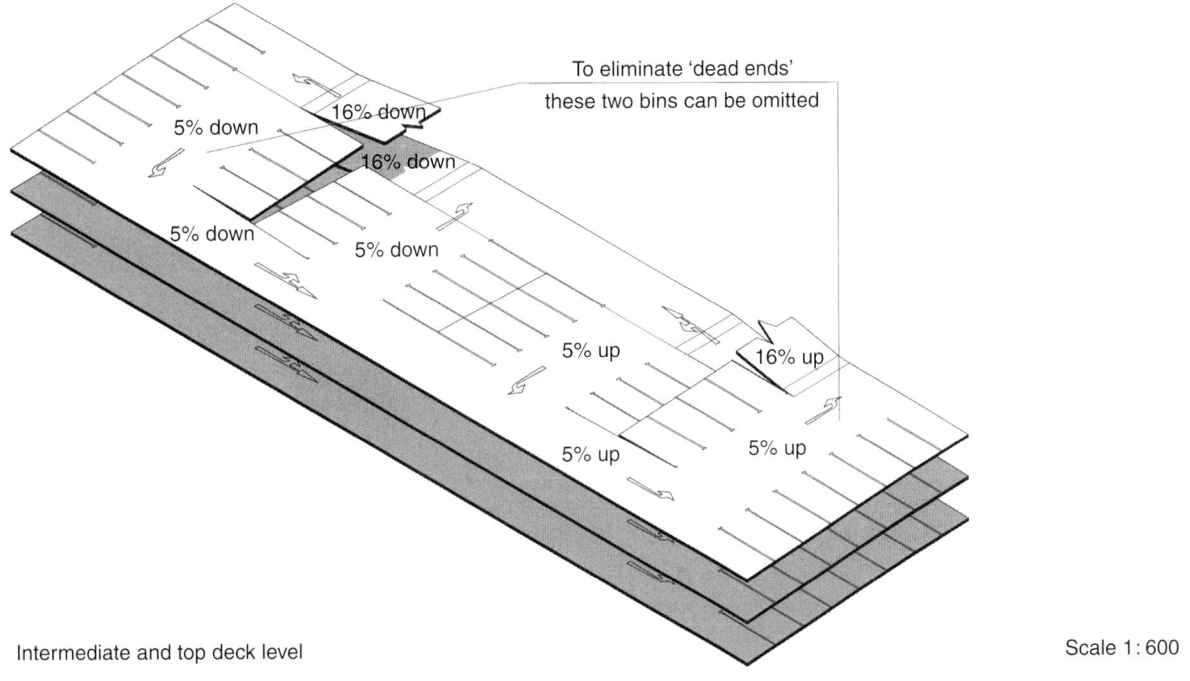

Intermediate and top deck level

Scale 1:600

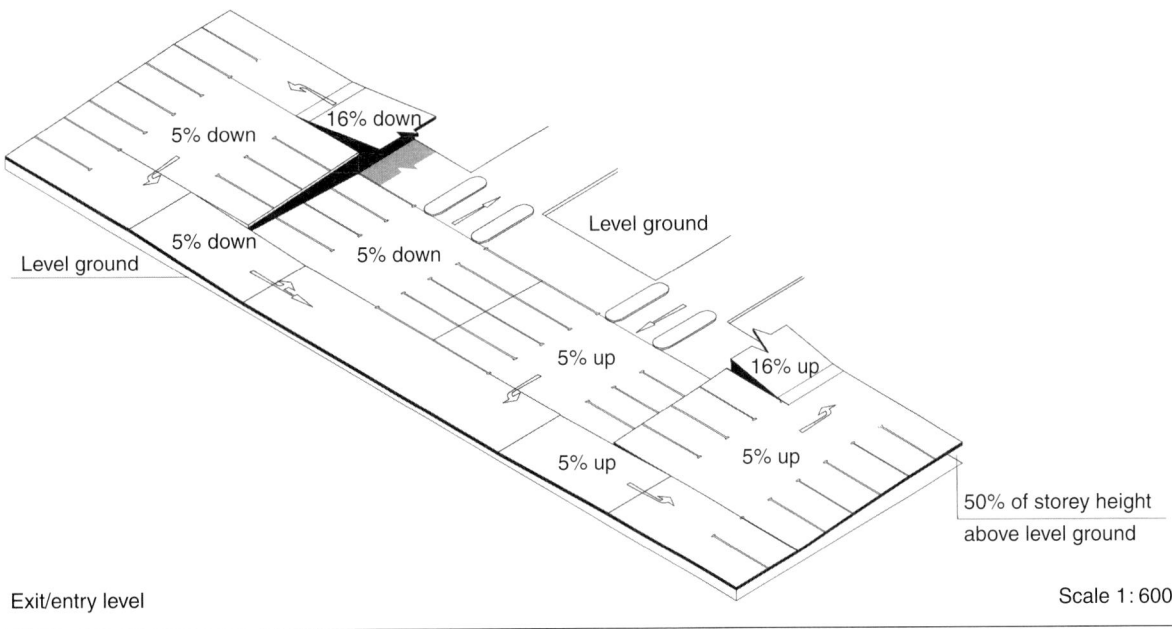

Exit/entry level

Scale 1:600

MD 5 (Figure MD 5)

ADVANTAGES

- It can be accommodated on sites down to 19.2 m in width, either above or below ground level.
- Each deck can be searched in succession, up or down.
- It has good re-circulation capabilities.
- A separate fire escape for the end bins is not required.

DISADVANTAGE

Its static efficiency is not good and actually reduces as the building increases in length.

COMMENTS

- This is a VCM variation of the MD 3 layout that enables pedestrians to gain access to all parts of a deck without encountering a slope in excess of 5%.
- Although shown as a four-bin-width minimum layout it can be constructed with three bin-widths where the central bin is a two-way flow.
- The high proportion of cross-ways to stalls for the internal bins makes for a reduced static efficiency when compared with long MD 1 and 2 layouts. However, this can be similar when the building length is six bin-widths or less (and benefits, too, from good re-circulation characteristics).

STATIC EFFICIENCY

Although the static efficiency is not good by normal standards, this layout can be constructed down to a length of three bin-widths, at which dimension comparison with other layouts is academic.

OTHER LAYOUTS

Apart from a MD 2B or MD 3 layout, there are no other layouts of similar dimensions with which comparisons can be made.

7.11. Circular sloping decks (CSD)

Two-way flow is the only practical circulation pattern for a car park layout based on a hollow circle or ellipse. One-way flow layouts consisting of two interconnected rings in a 'figure of eight' pattern have been proposed but are not considered to be practical and as such have not been featured.

Stand-alone circular car parks can only be justified, realistically, on architectural grounds. The constant turning, the 'follow my leader' inflow and outflow route, the driver's inability to see any reasonable distance ahead and the lack of a rapid outflow route, renders them less popular with motorists than most other circulation types that are available to designers. Effective security surveillance, also, is rendered more difficult when the parking is on a constant curve.

The smallest practical diameter with cars parked only on the outside of the aisle is about 34 m and can accommodate some 30 spaces for every 360° of rotation. Static efficiency is poor, being about 28 m^2 per space, but such layouts can be justified when stalls are added to the outside of a two-way-flow circular ramp that would have been there in any case. As the diameter increases, static efficiency improves and above 50 m it becomes quite good, but then the hole in the middle also gets larger and generally becomes wasted site space.

When the overall diameter reaches 65 m, the central hole becomes large enough to contain another circular car park. Both can then become one-way-traffic flows. However, the circulation pattern requires all traffic to drive up to the top parking level in order to join the outflow route. The awkward circulation, coupled with the large diameter required, also renders this an impractical proposition for the UK and as such, it has not been featured in this book.

Almost any site capable of incorporating a circular car park could also contain a rectangular layout that would be more 'user friendly' and have improved static and dynamic characteristics. Above ground level it cannot be justified on practical grounds alone.

When located underground, circular layouts can be justified on engineering grounds, as the drum shape is an ideal form to resist compressive forces exerted by the surrounding ground and the central void is available for pedestrian access and mechanical ventilation. However, a well-designed rectangular layout of the same parking capacity will still be more economical to construct, and will be more 'user friendly'.

Figure CSD 1 Circular sloping decks

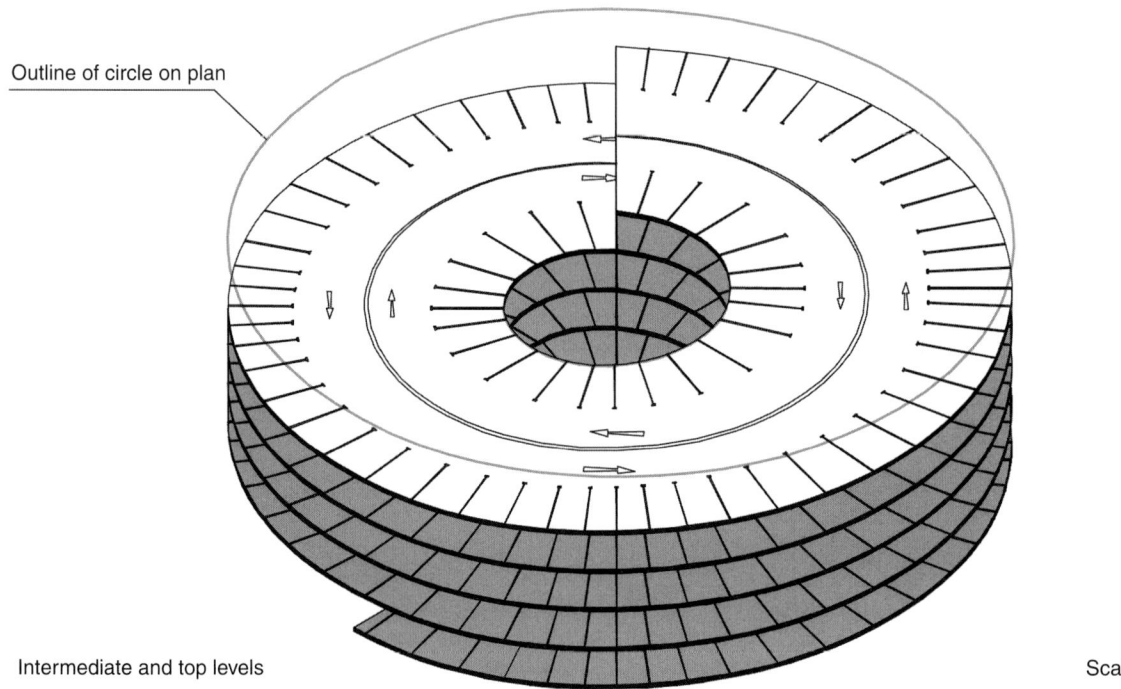

Intermediate and top levels — Scale 1:600

Static efficiency for different diameters

Diameter	Capacity inner ring	Capacity outer ring	Total	Area per car space	
35.000 m	N/A	33	33	25.00 m	(6.000 m wide one-way flow ramp)
40.000 m	N/A	40	40	26.40 m	(7.200 m wide two-way flow ramp)
45.000 m	15	46	61	24.40 m	
50.000 m	21	53	74	23.68 m	
55.000 m	28	60	88	22.90 m	
60.000 m	34	66	100	22.70 m	

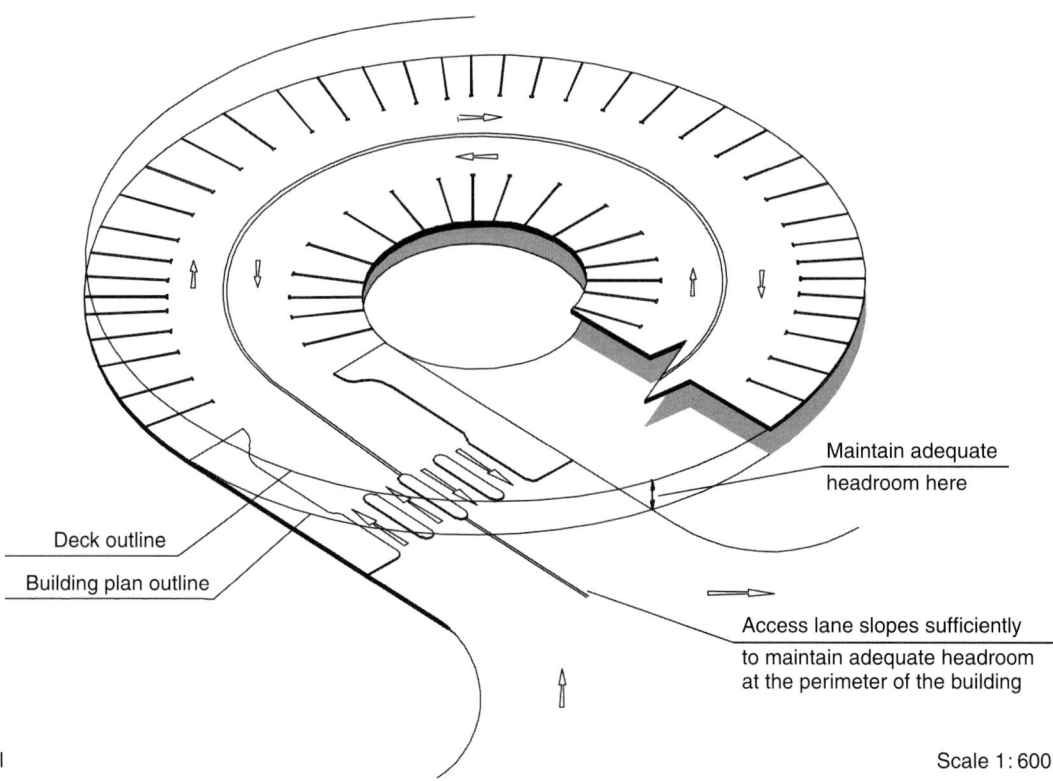

Exit/entry level — Scale 1:600

CSD 1 (Figure CSD 1)

ADVANTAGES
- All of the stalls are passed on the inflow route.
- The cylindrical shape is well suited to resist compression forces imposed by the ground when used in underground facilities.
- Good static efficiency for the larger diameter facilities.
- The sloping decks provide an unmistakable indication of the direction of traffic flow, the inflow route is upwards and the outflow route is downwards (reversed in underground facilities).

DISADVANTAGES
- All of the stalls are passed on the main outflow route.
- No natural recirculation capability.
- The need to turn constantly and the inability to see well ahead is an unpopular feature of the design.
- The circular layout renders CCTV supervision less effective than one with straight traffic aisles.

COMMENTS
- On anything other than a circular site, it is difficult to justify this layout above ground for reasons other than architectural. As the diameter increases so does the wasted site area created by the hole in the middle.
- It can be justified when it also doubles up as a spiral ramp providing access to a high-level parking area (with parking on one or more sides), but, even so, care has to be taken to ensure that the dynamic capacity of the two-way-flow aisle is not exceeded.
- The circular layout makes security supervision more difficult. Staff cannot see as far as when the aisles are straight and CCTV cameras will be required more frequently.

STATIC EFFICIENCY
- The number of stalls varies, dependent upon the diameter. At the smallest practical diameter of 40 m the area per vehicle space requirement is 26.4 m^2 but at a diameter of 60 m the area per vehicle space requirement reduces to 22 m^2 approximately: this can be deemed, Good, but then the hole in the middle gets larger and becomes wasted site space

OTHER LAYOUTS
- There is no alternative layout that embodies a circular shape, but on any site larger than, say, 32 m^2, a rectangular car park would produce a more economical and efficient layout.

7.12. Half-external ramps (HER)

When semi-circular ramps are used, the connection with a flat parking aisle at each deck level reduces the sense of unease developed by some drivers when they rotate through 360° or more. Consequently, HER ramp systems are quite popular with the parking public. In rectangular formats, they behave in the same manner as the internal ramping in SLD- or VCM-type layouts. Although shown with linked ramps, they can be separated and located on any wall face to best suit the highway access requirements.

When leaving a right-turning ramp and entering a traffic aisle, drivers need to be aware of traffic approaching from the left. The presence of passengers, obstructing sideways vision, could well render this a little awkward. When turning to the left, drivers can observe traffic approaching from the right without front-seat passengers obstructing the view. There are no absolutes in this consideration and often the direction of circulation is pre-determined by other factors, such as the need to exit the facility without crossing the path of entering traffic.

When compared to car parks with internal ramps the extra cost of a half-external ramp structure can be offset, to some extent, by the reduction in stalls used to complete the circulation route. When used in small parking facilities they can be uneconomic and site-area consuming, but as the capacity of the car park increases this factor becomes less significant. Often, the shape of the site enables HER ramps to be introduced without affecting the area available for parking.

The lowest levels on any outflow circulation route tend to become congested first, since traffic from the upper levels is incremental as it passes through the lower floors on its way out of the building. When the anticipated flow rate exceeds about 1300 cars per hour, a second ramp system or a fully external ramp access should be considered.

In arriving at a decision between the ramp types, it should be appreciated that the geometry of HER ramps enables motorists to re-circulate, rapidly, throughout all of the parking levels. However, being combined in part with the parking decks, their dynamic capacity will be a little less than that for ER-type ramps.

Circulation layouts

Figure HER 1 Half-circular ramps

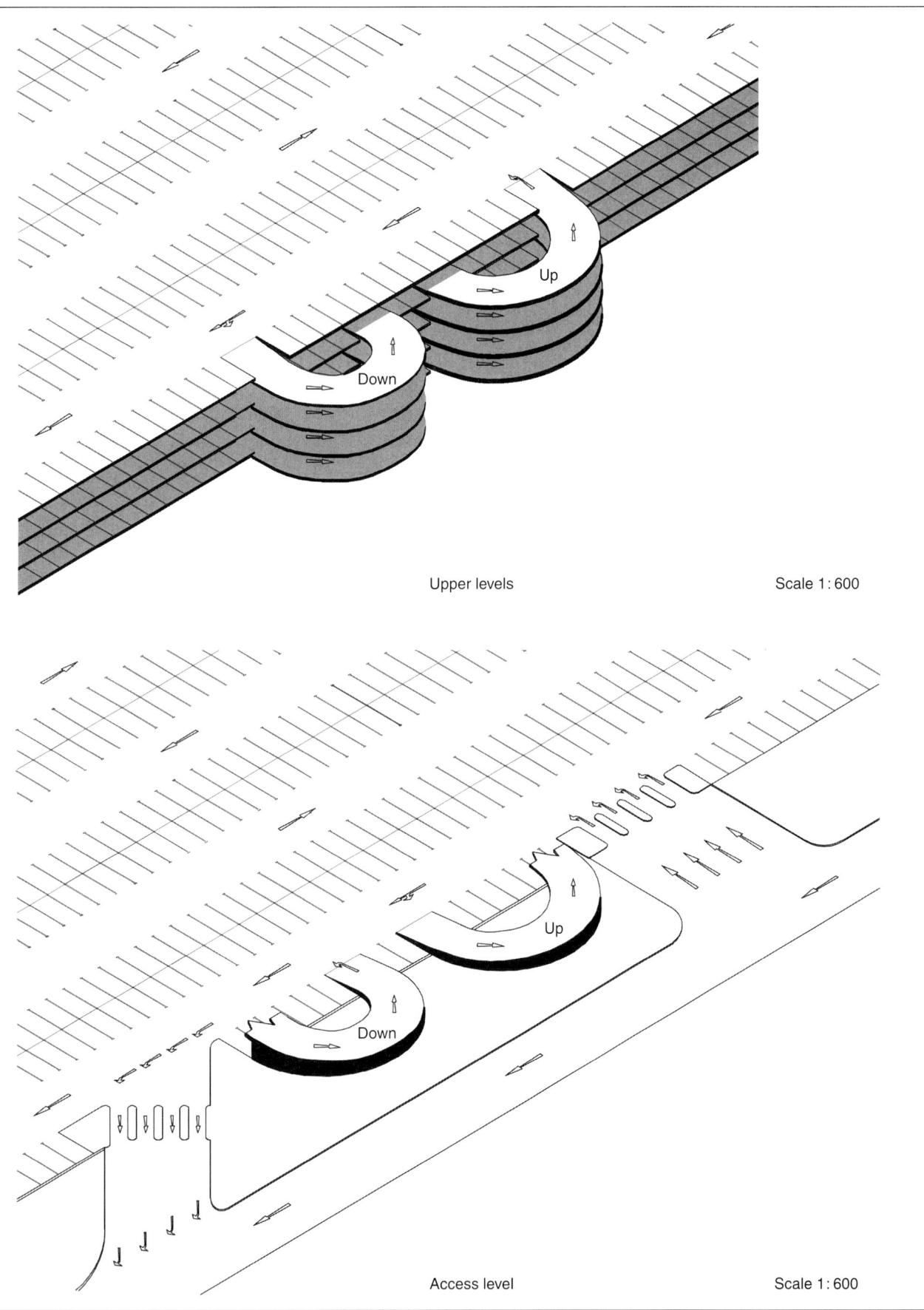

Upper levels — Scale 1:600

Access level — Scale 1:600

HER 1 (Figure HER 1)
ADVANTAGES
- Each vertical flow route only requires four stall spaces per deck to complete the vertical circulation.
- Each vertical flow route is rapid, passing only 12 stall spaces per deck.
- Each deck is completely flat and can be searched in succession, up or down with good re-circulation capabilities.
- Although shown adjacent to each other, they can be conjoined (see HER 3) or located separately on any face of the building to suit access and egress requirements.

DISADVANTAGES
- The ramps project 9.2 m from the face of the building,
- Both flow routes combine with the aisle traffic at each deck level. This, however, is not normally a significant matter.

COMMENTS
- As shown, when measured on the centre-line, the going is 32.2 m. For a storey height of 3 m this produces a continuous gradient of 9.3%. For storey heights over 3 m it may be necessary to extend the projection of the ramps beyond the face of the building, or alternatively, make them elliptical.
- A VMS located at the entrance to each deck will improve dynamic efficiency by eliminating the need to search, needlessly, any particular deck.
- When the ramps are conjoined, the re-circulation characteristics are enhanced.
- It is suitable for all usage categories where the dynamic capacity of the 'combined' section of an aisle is not exceeded (about 1100 vehicles per hour).
- When used for small, urban centre layouts the extra space requirements and the cost of the ramps may make them an uneconomical proposition.

STATIC EFFICIENCY
The static efficiency is totally dependent on the layout design of the parking deck.

OTHER LAYOUTS
Any of the featured ramp systems will perform equally as well; it is more a matter of choice rather than any other factor.

Figure HER 2 Three-slope ramps

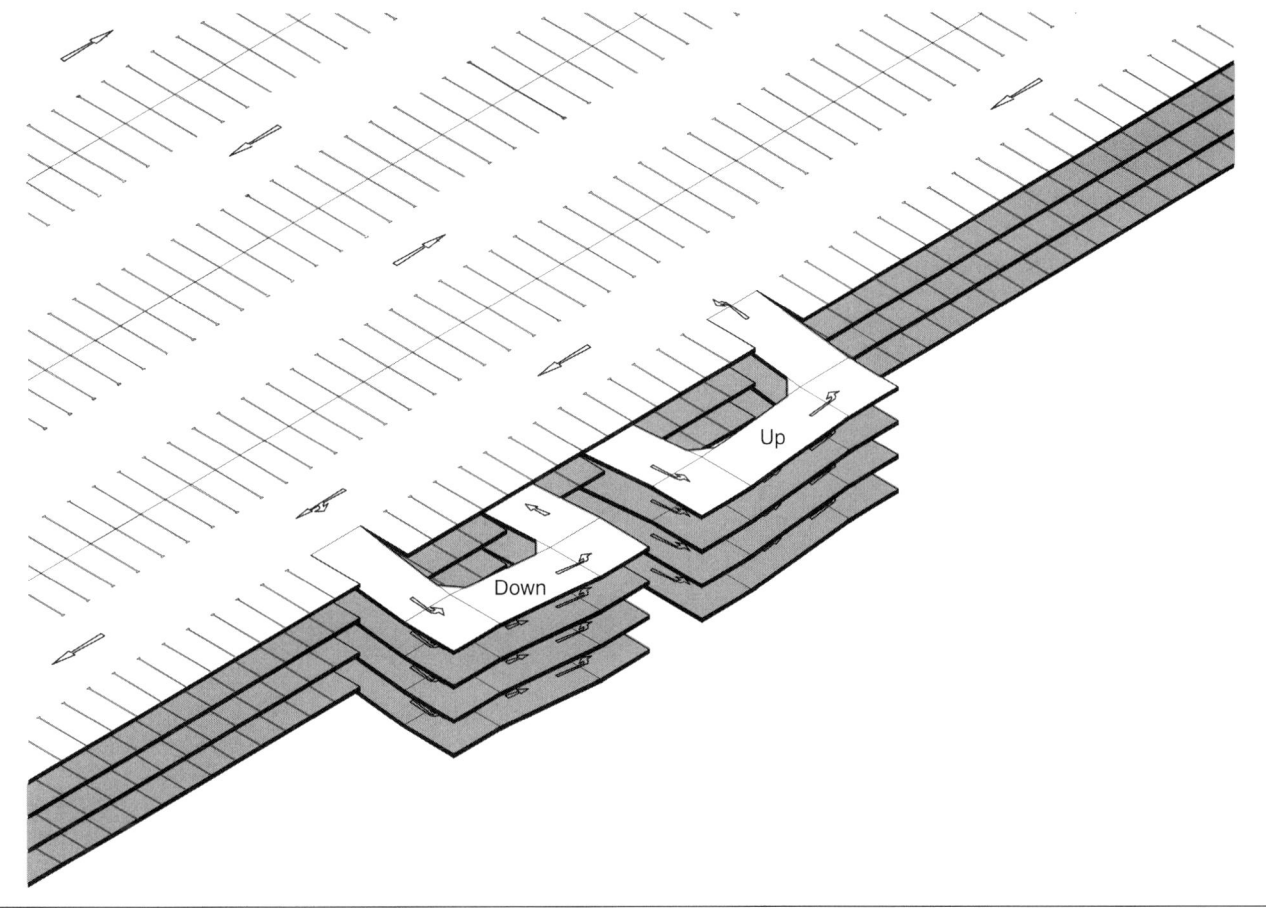

HER 2 (Figure HER 2)
This is a three-ramp rectangular system and is shown simply in order to visually compare it with the appearance of the HER 1 semi-circular ramp layout. In all other respects it is similar.

Figure HER 3 SLD-type ramps

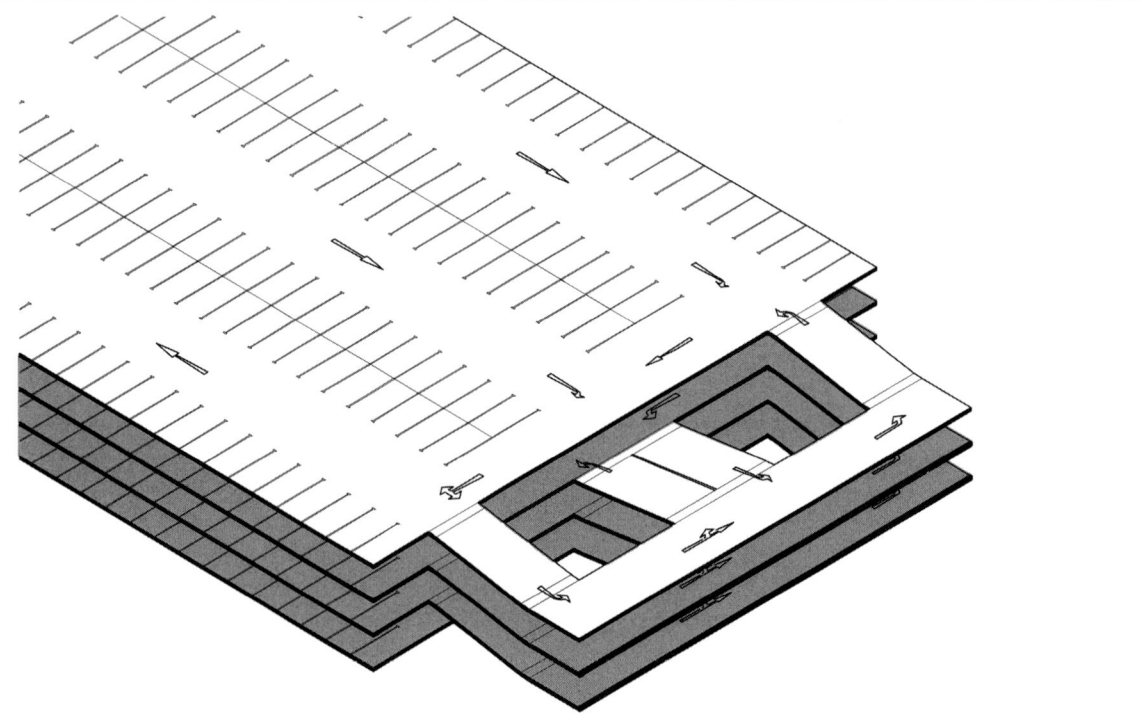

HER 3 (Figure HER 3)
- HER 3 is shown as a conjoined ramp system located at the end of a building where there is insufficient width for side mounting a HER 1 or 2-type layout.
- When the entry and exit points are located on opposite sides, access to from the upper levels can be rapid.
- In all other respects it functions the same as the others in the MD series.

7.13. External ramps (ER)

ER ramps function independently of the car parks they serve and their only contact is the access-ways into and out of the parking decks.

They are used mainly in large-capacity facilities and a single-ramp system can cope with 1500 stalls, approximately, for Cat. 1 and 2 layouts: 2000 for Cat. 4 and as many as 5000 stalls in Cat. 3 layouts.

The capacity of ramps that conform to the 'recommendations' can be assessed as 1500 vehicle movements per hour.

Although the one-way-flow ramps have been shown linked they can be separated and located on any elevation to best suit the site requirements.

As with circular car parks, circular ramp systems are less popular with the public than straight ramps, due, quite possibly, to the fact that designers have tended to underestimate the minimum diameter needed to create an acceptable driving environment. The fact that vehicles can turn more tightly than the diameters recommended for good practice does not mean that drivers are prepared to accept the situation.

The 'recommended' minimum diameters are for general acceptance by the public, not just for the vehicles they drive.

Figure ER 1 Circular ramp with two-way flow

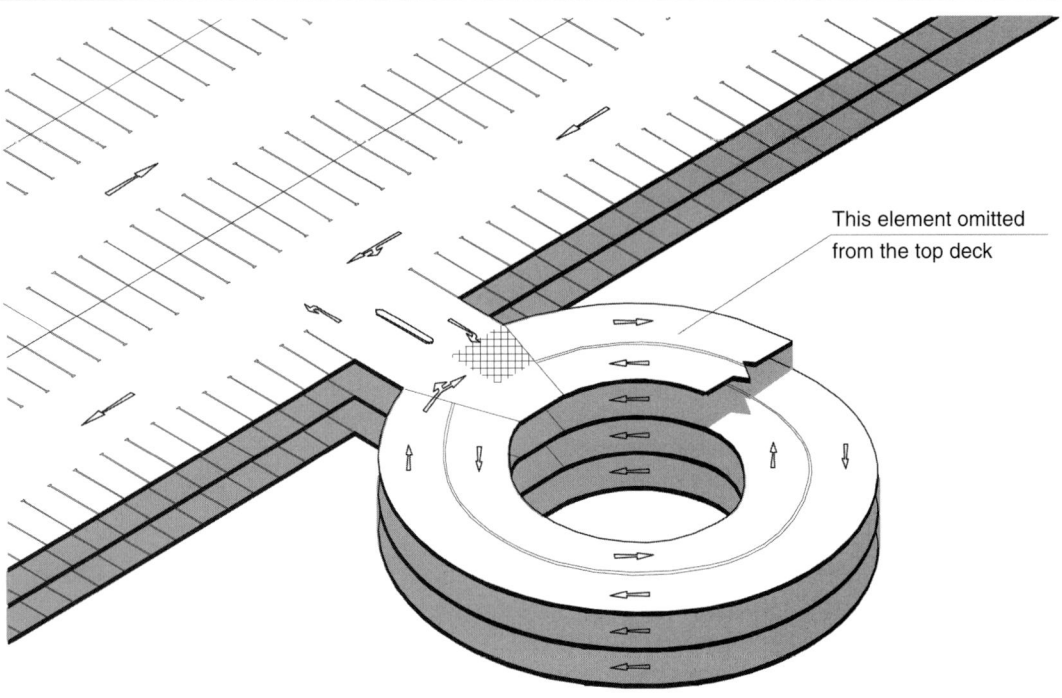

Intermediate and top deck level

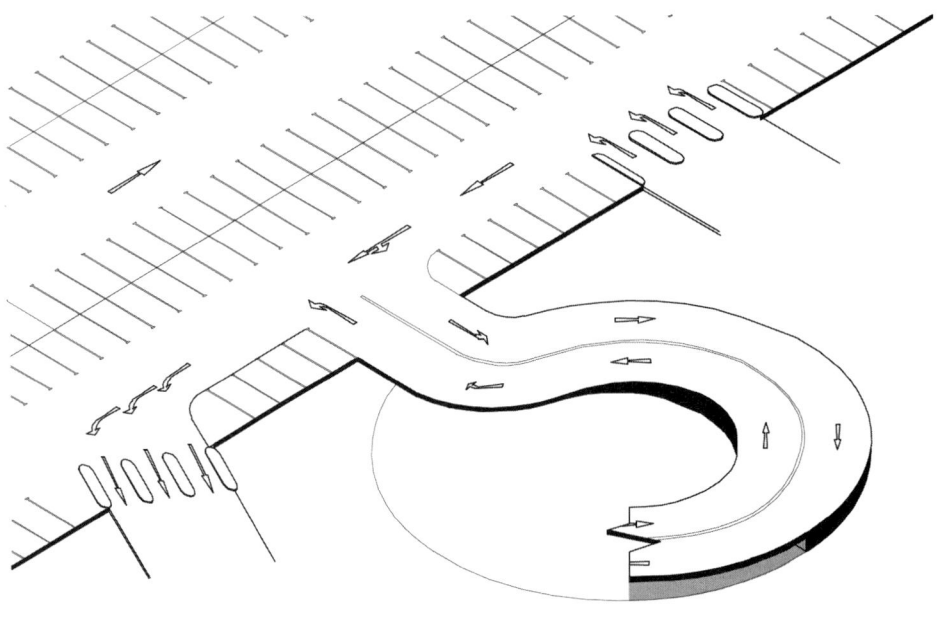

Exit/entry level

ER 1 (Figure ER 1)

ADVANTAGES

- Both flow routes are contained in a single structure, which could prove beneficial in site utilisation and construction costs.
- Four stall spaces per deck are required to complete the circulation route.
- The vertical circulation routes are unobstructed by other traffic, except when joining or leaving the ramp system, resulting in a dynamic capacity on the ramps of 1480 vehicles per hour.

DISADVANTAGES

- Traffic exiting the deck must cross the path of ramp traffic climbing to an upper level.
- The ability to observe approaching traffic on the opposite lane is limited by the ramp curvature.
- They are not a popular format with the parking public at the best of times and rapidly become more unpopular if the diameter reduces below the 'preferred' minimum.

COMMENTS

- Unless appearance or site utilisation considerations are pre-eminent, there is no advantage in choosing this format over other external ramp types.
- Providing that traffic drives on the 'correct' side of the aisle and the lanes are of an adequate width, a lane-dividing kerb is not considered to be an essential feature: double yellow lines should suffice.
- A variable message sign system, located at the head of the inflow ramp at the approach to each parking deck, will considerably improve dynamic efficiency by eliminating the need to search, needlessly, any particular floor.
- Suitable for all car park usage categories where the collective capacity of the exiting traffic at the lowest level is not exceeded.
- The capacity of the exit barriers and external road system requires careful consideration if traffic is not to 'back up' within the car park.

OTHER LAYOUTS

An ER 2 ramp system is a logical option and projects less distance from the parking deck.

Figure ER 2 Circular ramps with one-way flow

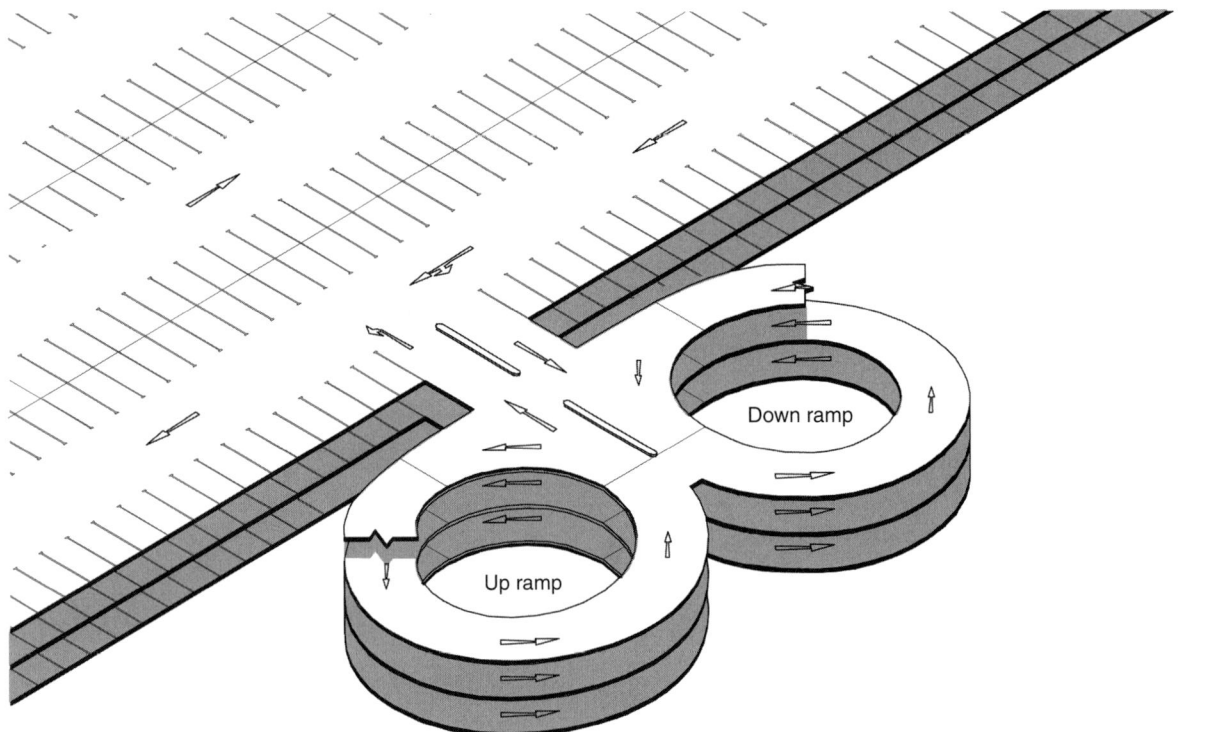

Intermediate and top deck level

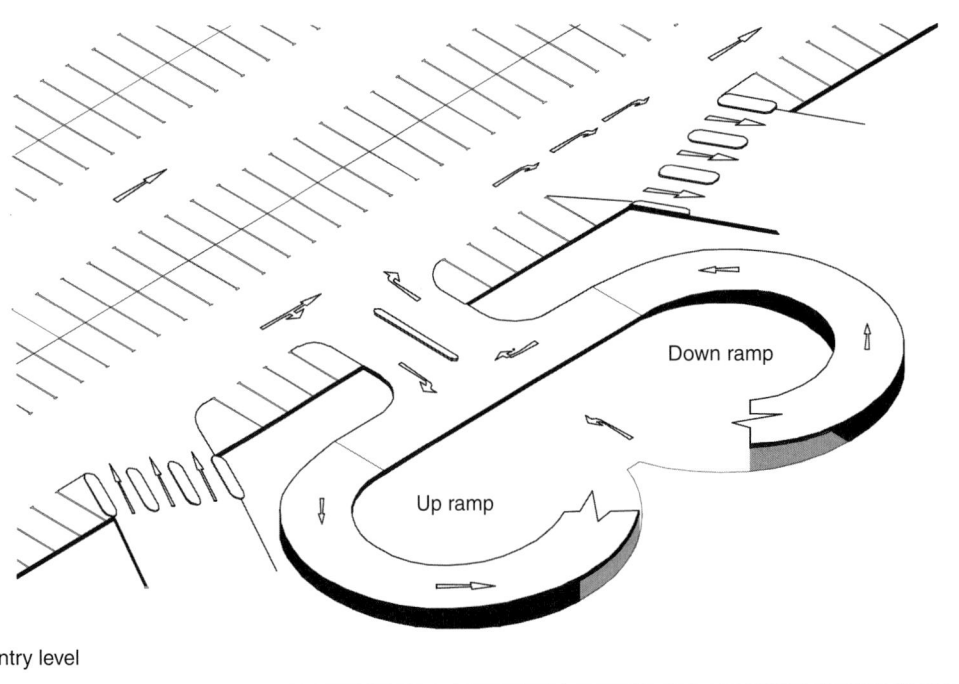

Exit/entry level

ER 2 (Figure ER 2)
ADVANTAGES
- Both flow routes can be conjoined or separated. They can be positioned at any convenient location around the car park.
- Four stall spaces per deck are required to complete the circulation route.
- The vertical circulation routes are unobstructed by other traffic, except when joining or leaving the ramp system, resulting in a dynamic capacity on the ramps of 1480 vehicles per hour.

DISADVANTAGE

They are not a popular format with the parking public at the best of times and rapidly become more unpopular if the diameter reduces below the preferred minimum.

COMMENTS
- Unless appearance or site utilisation considerations are pre-eminent, there is no advantage in choosing this format over rectangular external ramp types.
- A variable message sign system, located at the head of the inflow ramp at the approach to each parking deck, will considerably improve dynamic efficiency by eliminating the need to search, needlessly, any particular floor.
- Suitable for all car park usage categories where the collective capacity of the exiting traffic at the lowest level is not exceeded.
- The capacity of the exit barriers and external road system requires careful consideration if traffic is not to back up within the car park.

OTHER LAYOUTS
- An ER 1 ramp system is a logical option but the cross-over condition at the exit from each floor level renders it less efficient, dynamically.

Figure ER 3 Three-slope ramps with one-way flow

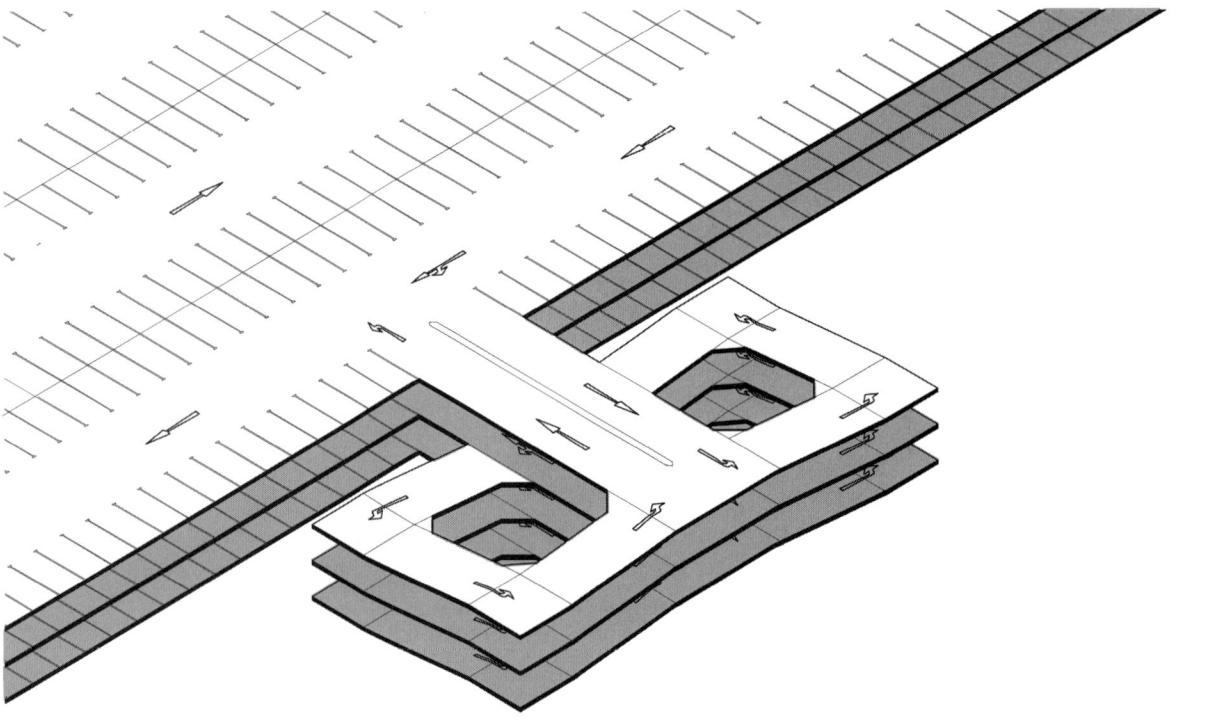

Intermediate and top deck level

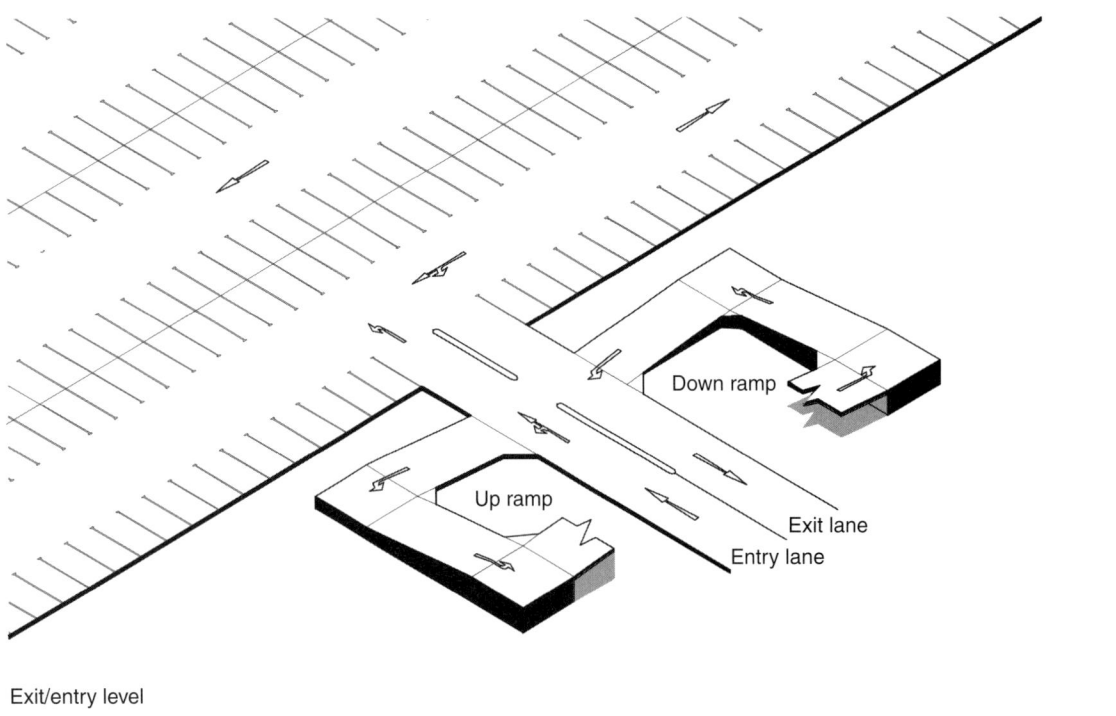

Exit/entry level

ER 3 (Figure ER 3)

ADVANTAGES
- Both flow routes can be conjoined or separated. They can be positioned at any convenient location around the car park.
- Four stall spaces per deck are required to complete the circulation route.
- The vertical circulation routes are unobstructed by other traffic, except when joining or leaving the ramp system, resulting in a dynamic capacity on the ramps of about 1500 vehicles per hour.

DISADVANTAGE

The projection from the side of the car park is some 10 m greater than for a similar HER-type ramping system.

COMMENTS
- Shown with split levels for demonstration purposes, it is equally adaptable to the other systems featured in the HER series.
- A variable message sign system, located at the head of the inflow ramp at the approach to each parking deck, will considerably improve dynamic efficiency by eliminating the need to search, needlessly, any particular floor.
- Suitable for all car park usage categories where the collective capacity of the exiting traffic at the lowest level is not exceeded.
- The capacity of the exit barriers and external road system requires careful consideration if traffic is not to back up within the car park.

OTHER LAYOUTS

Layouts utilising three-ramp- or VCM-type systems could be used to similar effect.

Figure ER 4 External express ramps

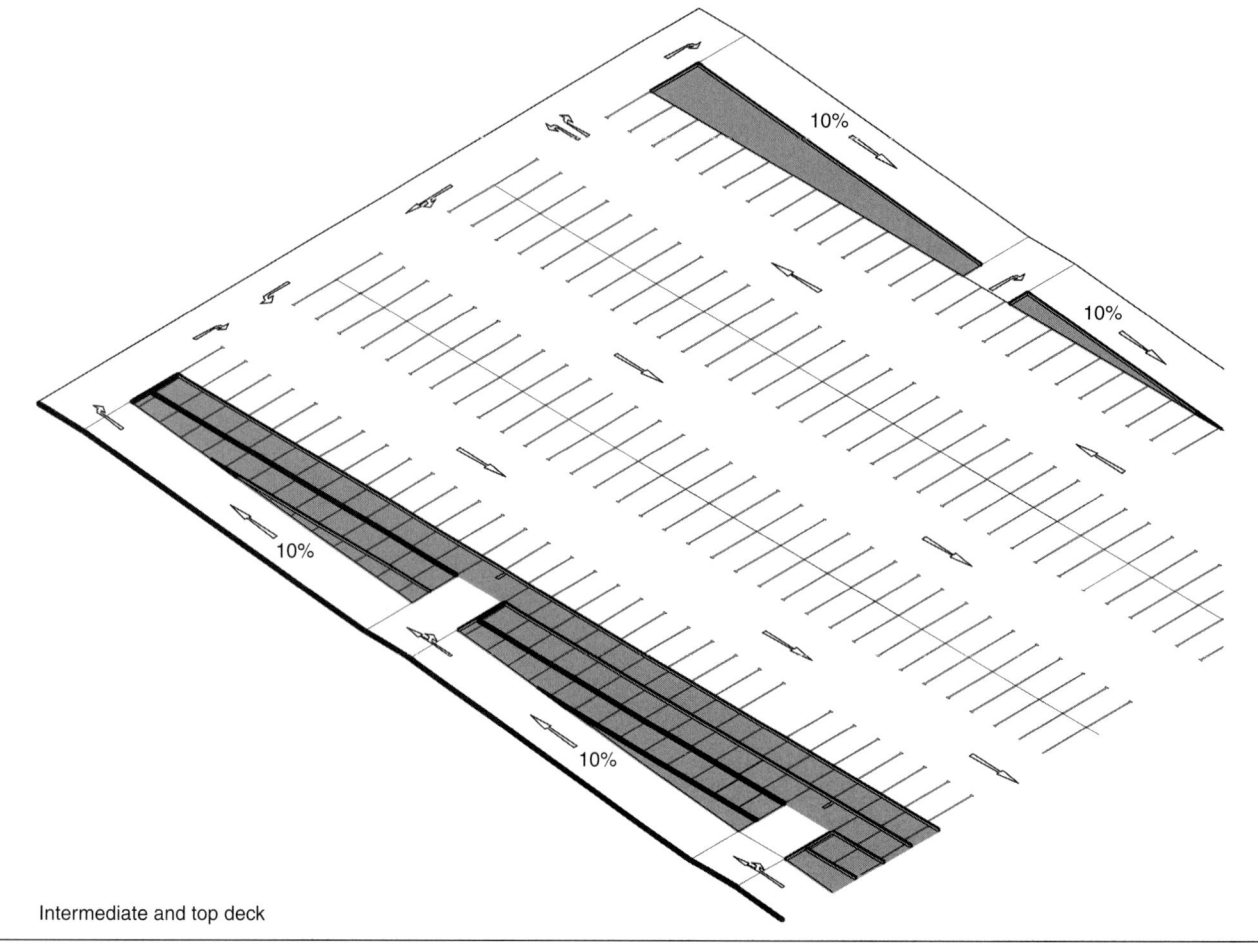

Intermediate and top deck

ER 4 (Figure ER 4)

ADVANTAGES
- Four stall spaces per deck are required to complete the circulation route.
- The vertical circulation routes are unobstructed by other traffic, except when joining or leaving the ramp system, resulting in a dynamic capacity on the ramps of 1480 vehicles per hour.
- The ramps can be combined on one side of the car park and that could save some 5 m in overall width requirements, but this will involve opposite direction traffic crossing paths.

DISADVANTAGE
For each level climbed or descended, the 'going', at 8.5%, is 14 stall-widths per 2.9 m storey height. The length of the site, therefore, will dictate the number of parking levels.

COMMENTS
- A variable message sign system, located at the head of the iflow ramp at the approach to each parking deck, will considerably improve dynamic efficiency by eliminating the need to search, needlessly, any particular floor.
- The projection from the side of the building can vary, but the turning dimension onto the parking deck must be considered.
- Suitable for all car park usage categories where the collective capacity of the exiting traffic at the lowest level is not exceeded.
- The capacity of the exit barriers and external road system requires careful consideration if traffic is not to back up within the car park.

OTHER LAYOUTS
There are no other external ramp systems that project as little from the sides of a building.

Figure ER 5 Stadium-shaped interlocking ramps

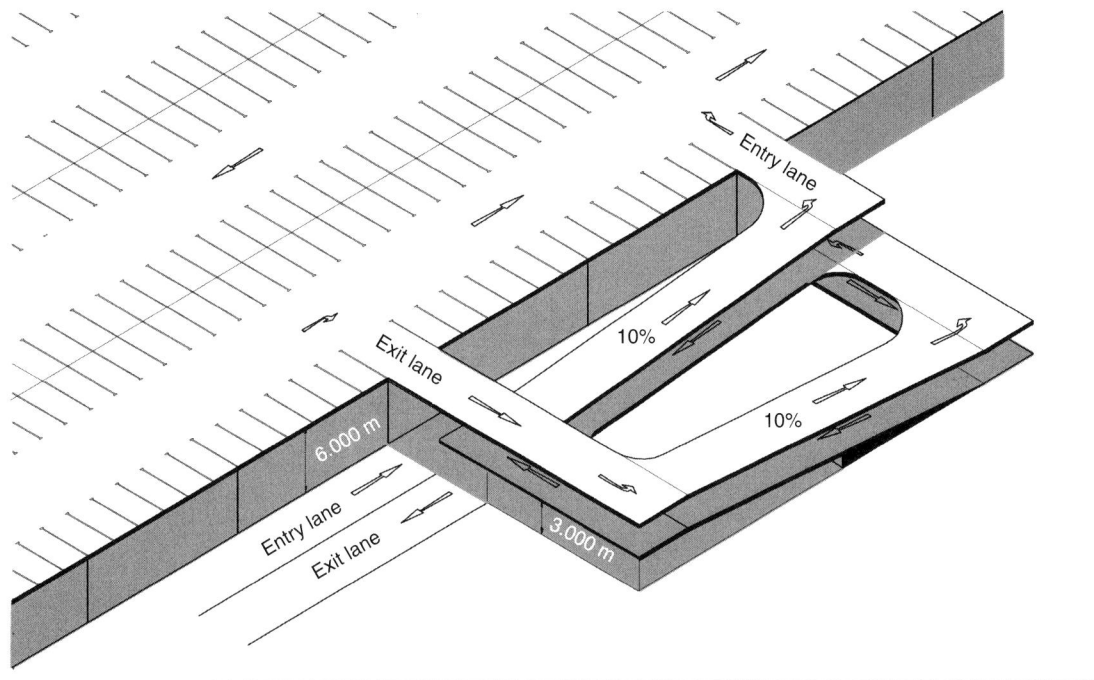

ER 5 (Figure ER 5)

ADVANTAGES
- The system is intended for use when a single-storey height of 5.4 m, or greater, is required. One of the flow routes can then be inserted between the other. This effects a reduction in the plan area when compared with other external ramp layouts where the flow routes are constructed side by side.
- The straight ramp element of the layout is less daunting to motorists than similarly dimensioned circular ramp systems.

DISADVANTAGES
- The entry and exit locations are not very flexible and could cause problems. To locate them side by side could involve increasing the site width, locally, at ground level.
- Locating the entry and exit control barriers at the first parking level may well be the best solution for large-capacity car parks.

COMMENTS
- The main purpose of this ramp system is to gain access to a car park located above a commercial or retail operation where an increased storey height is required.
- Ramp width dimensions as little as 16 m between kerbs have been noted without any apparent reduction in popularity, but, generally, the wider the better.
- It can be used to gain access to all of the parking decks, but it is to be remembered that with normal storey heights only alternate parking levels can be accessed on the way up.
- Suitable for all parking categories but another ramping system should be adopted where more than one level of parking occurs.
- Within the body of the parking area, a vertical circulation system that visits each level in turn will be preferable.
- An option is to carry on straight up but change to an HER system above the first parking deck level. The overall dimensions for the ER 5 and the two HER ramps would need to be compatible for the best structural efficiency.

OTHER LAYOUTS
There are no other layouts that can operate to as high a standard within the plan area of this system.

Figure ER 6 Circular interlocking ramps

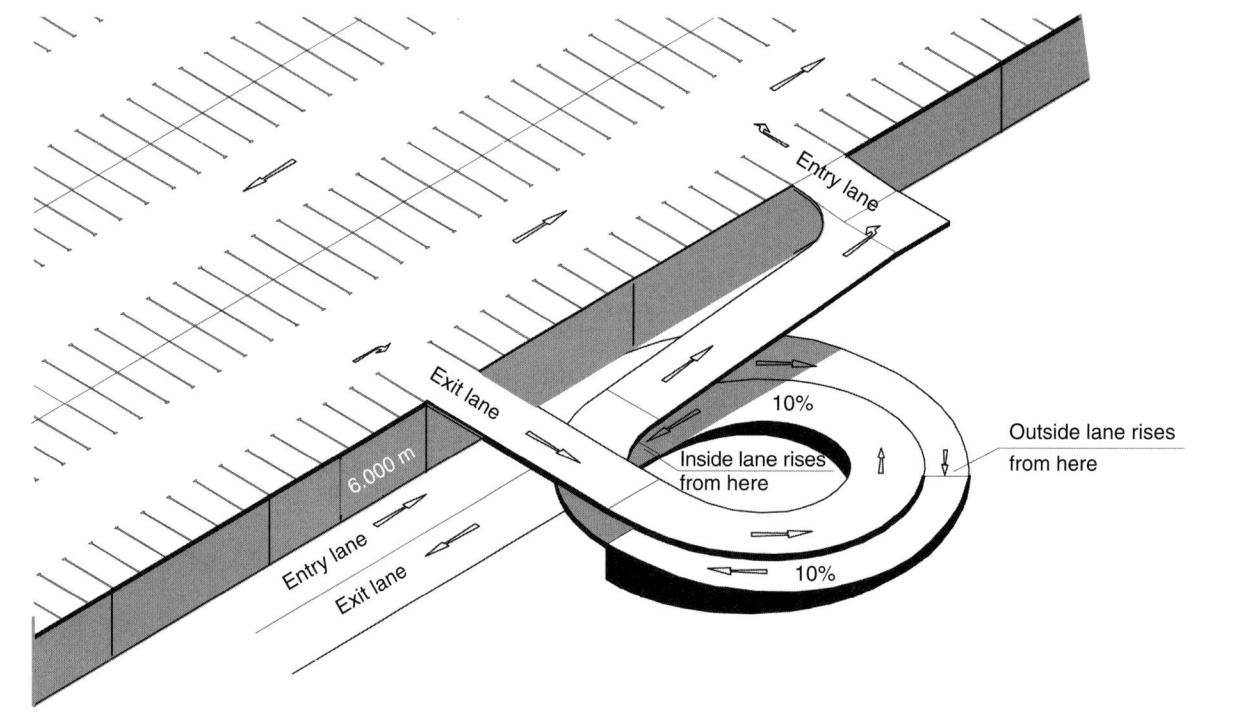

ER 6 (Figure ER 6)

ADVANTAGE

An alternative to ER 5, this system is also intended for use when a storey height of 5.4 m, or greater, is required. One of the flow routes can then be inserted between the other, thereby effecting a reduction in the plan area when compared with other, side-by-side, circular ramp layouts.

DISADVANTAGES

- The entry and exit locations are not very flexible and could cause problems. To locate them side by side could involve increasing the site width, locally, at ground level.
- For large-capacity car parks, where a number of entry and exit control barriers may be required. Locating them at the head of the ramps, in the body of the car park, may well be the only solution.

COMMENTS

- It operates in a similar fashion to an ER 5 ramp system and performs the same function.
- If the site widths are restricted then a stadium-shaped ramp could provide a better solution, but otherwise the choice of which ramp system to adopt is mainly one of client choice based on visual appearance.
- Measured on the ramp centre-line, a slope of 8.5%, for a 5.6 m-storey height produces a required going of 64.8 m. The minimum overall ramp diameter, therefore, will need to be in the order of 25 m.

ALTERNATIVE LAYOUTS

There are no other circular ramp layouts that can operate to as high a standard within the plan area of this system.

Car Park Designers' Handbook
ISBN 978-0-7277-5814-9

ICE Publishing: All rights reserved
http://dx.doi.org/10.1680/cpdh.58149.121

Chapter 8
Stairs and lifts

8.1. Discussion

The location of vertical services for pedestrians can have a significant bearing on the choice of vehicle circulation layout. If located at the end of a building, flat-end vehicle access-ways, also capable of being used by pedestrians, are preferable to ramps or steps between split-levels. When the vertical access is located along the building flanks, flat external deck layouts are preferable to those incorporating sloping decks (see Figure 8.1).

Unless otherwise agreed with the local Fire Officer, when used as fire escapes, the number and location of stairs should comply with the requirements of *Approved Document B* of the Building Regulations. Although not specifically referred to in Table 2 of that document, they are normally classed as Purpose Group 7 (*Storage and other non-residential*) for intermediate decks, or Purpose Group 2–7 (*Plant room or rooftop plant for top, exposed decks*). Where the top deck is roofed over this last requirement is not applicable. Extracts from the relevant clauses are shown in Section 8.2.

Figure 8.1 Birmingham International Airport MSCP, main atrium

The choice of whether or not to introduce lifts is one that has to be considered frequently. For Cats 1 and 3 use, lifts and/or a pedestrian ramp will be desirable even for a single suspended parking deck. For Cats 2 or 4 use, with two suspended parking floors, lifts are not essential provided that they are not required for transporting trolleys, luggage, disabled drivers, or carers with prams. For three or more suspended parking decks, lifts are recommended for all car park categories. The number and capacity of the lifts will vary according to the parking category and capacity.

8.2. Vertical and horizontal escape
8.2.1 Stairs: widths of flights

The Building Regulations provide basic rules for determining flight widths. *Approved Document B* states that, in the absence of other information on the likely number of occupants, guidance is given in Table C1. For car parks the factor given is two persons per parking space.

Use of this guidance tends to result in an overly conservative estimate on the building occupancy at any one time, for the following reasons.

- Vehicle occupancy in urban areas at peak periods has consistently been observed to be an average of no more than 1.4 persons per vehicle.
- Most of the time, parked cars are not occupied and so create no requirement for pedestrian fire escapes.

There are three basic occasions when car occupants and pedestrians will require access to a fire escape. These are

- when the vehicle enters the car park and is being driven around to find a parking space
- when pedestrians are walking to and from the deck exit location
- when the vehicle is being driven out of the car park.

In addition, there will be occasions when the car park staff are present, but their numbers are so small, relatively, that they can be discounted.

The maximum occupancy will relate to the peak traffic flows, both in and out. This information is usually available from the Traffic Assessment that would normally accompany the planning application for a new-build car park. In the absence of such information an assessment of the peak-hour traffic-flow can be taken as 65% to 70% of the total capacity; that is, for a 1000-space car park the peak flow over a one-hour period will be between 650 and 700 vehicles.

It is reasonable to assume a maximum parking period of five minutes for an efficiently designed layout (see Section 6.5) and another five minutes for pedestrians to exit the building: a total time of ten minutes. Usually the only people to exceed this time will be staff members and those very few who are waiting in the vehicle for others to return. Using the example of 700 vph, a ten-minute period results in 117 vehicles. Using, also, the Building Regulation figure of two persons per vehicle, the maximum total number of people present within the building over any ten-minute period will be 334.

In the event of a fire, when entering escape stairs without lobbies, it is possible for one flight of them, on any particular level, to be put out of action and this must be taken into account when calculating stair and door-opening widths. When lobbies are introduced all of the stair cores can be counted. Stairs without lobbies are easier to enter, especially for disabled pedestrians; they also take up less room, but an extra stair core is required.

From Table 7 of *Approved Document B*, it can be seen that, with three suspended levels and three escape stairs without lobbies, stair flight widths of 1000 mm will be adequate. For the main pedestrian stair core where there could be regular two-way pedestrian movements, a flight width of 1200 mm is recommended. Landings must be of a width no less than the flights they serve (see Figure 8.2).

Table 7 Vertical escape

Floor levels	Stair widths (from B1, Vertical escape)								
	1000 mm	1100 mm	1200 mm	1300 mm	1400 mm	1500 mm	1600 mm	1700 mm	1800 mm
1.	150	220	240	260	280	300	320	340	360
2.	190	260	285	310	335	360	385	410	435
3.	230	300	330	360	390	420	450	480	510
4.	270	340	375	410	445	480	515	550	585
5.	310	380	420	460	500	540	580	620	660
6.	350	420	465	510	555	600	645	690	735
7.	390	460	510	560	610	660	710	760	810
8.	430	500	555	610	665	720	775	830	885
9.	570	540	600	660	720	780	840	900	960
10.	510	580	645	710	775	840	905	970	1035

As an alternative to using Table 7, the capacity of stairs 1100 mm wide or wider (for simultaneous evacuation) can be derived from the formula:

$$P = 200W + 50(w - 0.3)(n - 1), \quad \text{or} \quad W = \frac{P + 15n - 15}{50 + 50n}$$

where P is the number of people that can be served; W is the width of the stairway, in metres, and N is the number of storeys served.

Notes
1 Separate calculations should be made for stairs/flights serving basement storeys and those serving upper storeys.
2 The population 'P' should be divided by the number of stairs available.

8.2.2 Horizontal escape

Widths of escape routes (from B1, Table 4)

Maximum number of persons	Minimum width
50	750 mm
110	850 mm
220	1050 mm
>220	5 mm per person

Notes
1 Widths less than 1050 mm should not be interpolated.
2 5 mm per person does not apply to openings serving fewer than 220 persons.

8.3. Escape distances

Approved document B specifies maximum pedestrian escape distances of the following.

- Intermediate decks – 25 m to an available escape in one direction only or 45 m to available escapes in two or more directions.
- Top, exposed decks – 60 m to an available escape in one direction only or 100 m to available escapes in two or more directions.

These escape distances can be increased by the use of a fire-engineered solution outlined in BS 9999 and discussed in further detail under Section 16.2 of this manual.

When measuring escape distances, it is not acceptable for the route to pass between parked cars unless a dedicated path is provided measuring no less than the width of the escape door. The door widths can be determined from the rules provided in the Building Regulations. It is to be appreciated that the doors on each level only need to contend with the number of vehicles on the deck they serve.

Figure 8.2 Escape stairs: typical dimensions

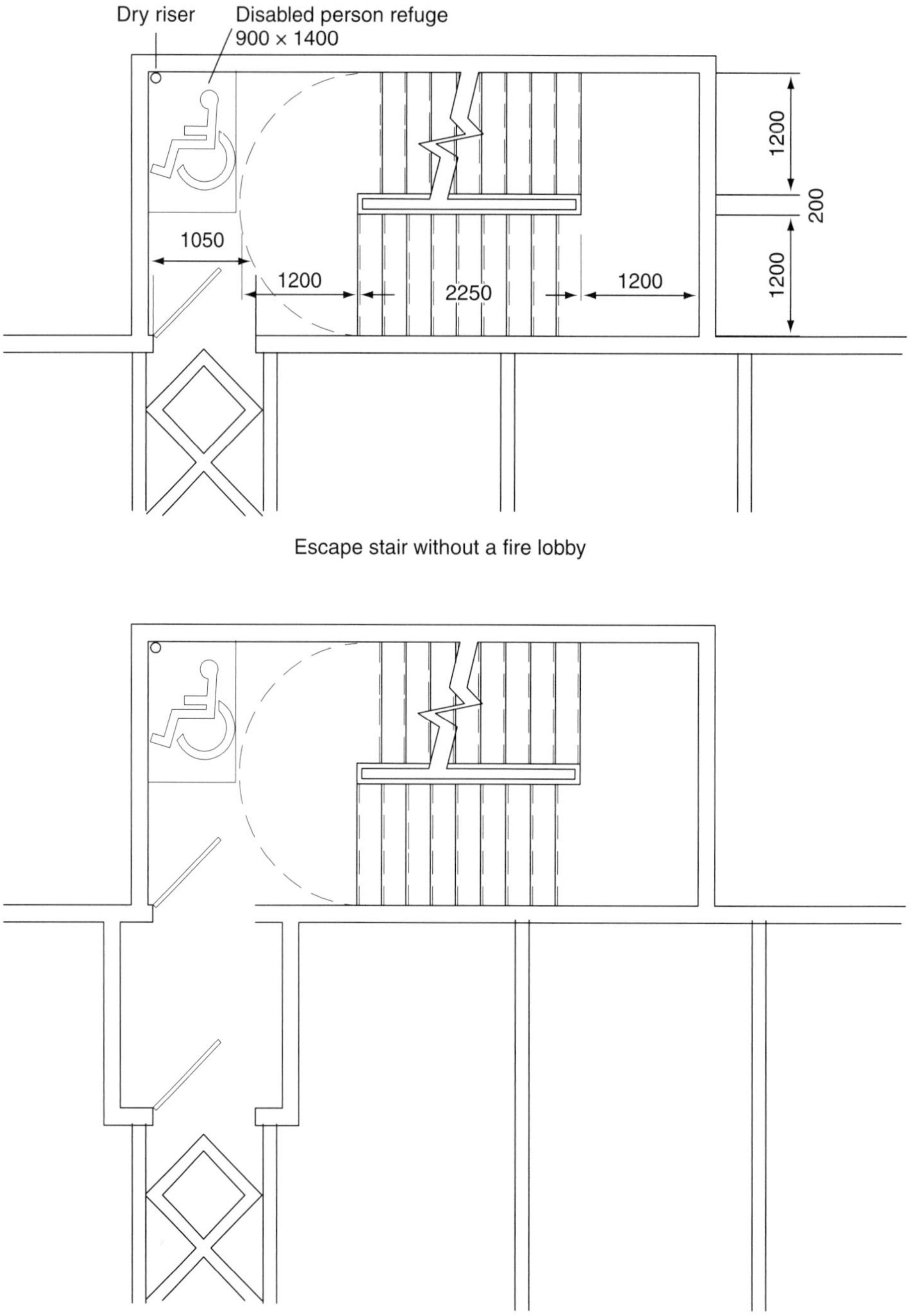

Escape stair without a fire lobby

Escape stair with a fire lobby

Where an escape route encounters a split level between adjacent parking decks, a ramp and/or steps may be required (see Figure 8.3). Ramps or stairs forming part of an escape route should comply with *Approved Document Part M, Access and facilities for disabled people*.

The final exit from the building should be to the open air. Ideally, the escape stairs should be located around the perimeter of the building. Should they be located internally then a fire-protected corridor must be provided that leads directly to the perimeter and the open air.

8.4. Lift sizing

Lift manufacturers usually provide a service for determining the number and sizing of lifts required for any particular building. For preliminary estimating purposes, however, a 'rule of thumb' method can be used for assessing the required lift capacity.

Figure 8.3 Pedestrian ramp layout between split-level decks

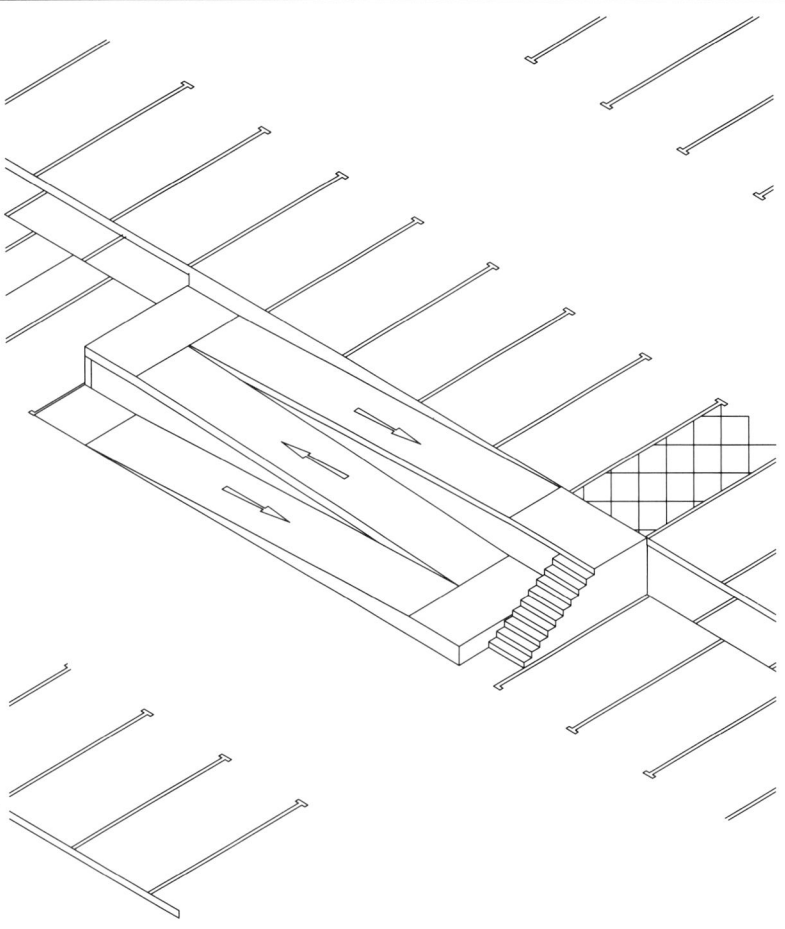

Combined stair and 5% slope, ramped access route between split-level decks. The ramps can be up to 1.600 m wide to enable prams and wheelchairs to pass each other and the stairs can be 1.200 m wide

For Cat. 1 parking located on a single level directly over a supermarket, 33-person lifts that can contain four shopping trolleys, double stacked, will produce an effective solution. Each supermarket chain has its own special requirements, but for preliminary design purposes where the end user has not yet been chosen and where all the parking is above ground, the lift capacity can be estimated at the rate of one 33-person lift for every 800 m^2 of retail shopping area.

The total parking capacity and ratio of suspended-level to surface-level parking will affect the lift requirement. Total parking capacity can be estimated at the rate of one parking space to every 8 m^2 of retail shopping area, and the lift requirement can be reduced proportionally to the ratio of suspended-level to surface-level parking.

Where more than one upper parking level is involved, double trolley-stacking is less effective than single trolley-stacking using 21-person lifts that are capable of containing two shopping trolleys each, side by side. The complete lift cycle will take longer when compared with single-level parking and will vary according to the number of parking levels to be served, but for preliminary design purposes they can be estimated at the rate of one 21-person lift for every 300 m^2 of retail shopping area.

Other Cat. 1 and 2 parking, not requiring shopping trolleys, can be estimated using a lift-space-per-person ratio of one for every 15 parking stalls on the suspended levels. For a 600-place facility with five suspended levels, 33 spaces will be required: say, three lifts of 13-person-, or four lifts of eight-person-capacity.

Cat. 3 parking can vary dramatically, from light use where one lift space per 25 'served' vehicle spaces can be adequate for much of the time, interspersed with periods of extreme activity when motorists drive to an airport to pick up, say, some 400 passengers arriving on a particular long-haul flight. The lift capacities can be assessed in a manner similar to the one described for supermarket use. However, a minimum of 25 trolley spaces should be provided for each major passenger aircraft arrival per hour. In airport-type situations, when it is known that future developments will occur such as the Airbus A380, which is capable of carrying more than 700 people, it is preferable to over-provide the lift services rather than under-provide.

Cat. 4 parking, unless there is information to the contrary, can be treated in similar fashion to Cat. 2. Although the arrival and departure times are more intense, there will be less luggage, prams and trolleys and more people will use the stairs if the lift capacity is over-extended.

Unless the car park is very small, two lifts should be considered as the minimum in order to contend with a breakdown or maintenance 'down time' on one of them.

Lift doors must be wide enough for persons in wheelchairs, and double pushchairs, to enter.

Worked example for a Cat. 1, 1000-space facility on four suspended deck levels (200 spaces per floor) with the main entry and exit at ground level.

The Traffic Assessment estimates an a.m. peak of 692 vph and a p.m. peak of 606 vph.

Ground-level parking will be the most popular (usual scenario from observations) followed in reducing popularity by the upper levels.

Assume 4×17-person lifts operating as duplex units, in two stair/lift towers and serving alternate floor levels; that is, G1 and G3 on one side and G2 and G4 on the other. The storey height is 2.8 m (note that the practical lift capacity has been reduced to 14 persons).

Each complete cycle for a pair of lifts averages $14 \times 2.8 = 39.2$ m travel distance

Lift speed = 1.6 m/sec

Time to open and close doors = 10 secs

Time to load/unload passengers = 10 secs

Note that at peak times each lift will stop three times per cycle.

The total journey time for each complete cycle, per pair, will be

Travel time	39.2/1.6 =	25 secs
Door time	6 × 10 =	60 secs
Passenger movements	6 × 10 =	60 secs
Total time		145 secs

Lift capacity per pair = 14 persons × 2 × 3600/145 = 694 persons per hour.

Two banks of lifts, therefore, provide a total capacity of 1388 persons per hour.

With an average of two persons per vehicle, the *Approved Document* requirement is 692 × 2 = 1384. At the more realistic figure of 1.4 persons per vehicle the requirement will be reduced to 969 persons, implying that the lift facility will be operating well within its capacity. If occasional short periods of extended waiting time can be tolerated,

4 × 13-person lifts can also provide an adequate capacity. It should be appreciated, however, that maintenance and breakdowns will occasionally reduce the number to three, and 17-person lifts can cope better when such a situation occurs.

It should be noted that calculations can only produce generalisations and the answers should not be considered to be precise figures.

Car Park Designers' Handbook
ISBN 978-0-7277-5814-9

ICE Publishing: All rights reserved
http://dx.doi.org/10.1680/cpdh.58149.129

Chapter 9
Disabled drivers and access assistants

9.1. Discussion

Although the necessary number of stalls and their dimensions are well documented, little comment has been made about the disposition of stalls for disabled drivers within a car park. The required provision for general public parking is up to 6% of capacity and subject to planning guidelines. This figure varies according to the use for which the car park is intended, but many local authorities now insist on 6% regardless of the car park's function. Some car parks have stalls spread around in ones and twos on a number of deck levels, wherever there has been an extra-width stall available. This is a cynical solution to the requirements. It ignores the particular mobility problems of some disabled people and is at odds with current Disability Discrimination Act recommendations. There is also a need to consider parking provision for a proportion of 'high top disabled' vehicles requiring 2.6 m clearance. It is acceptable to provide external parking bays close to the facility but if that is not possible, the ground parking level will have to be used. The interior should be so designed that the increased headroom occurs only locally and the vehicles can exit the facility without having to be driven through the whole of the ground parking level. It is unreasonable to have to increase all of the ground floor headroom just for a few parking bays. Statistically, the high bay vehicles form a small percentage of overall blue badge holders and are thus reflected in the allowable provision (see Figure 9.1).

The requirements for pedestrian ramps are based on places of work and residence wherein disabled persons can spend many hours per day. In such cases, it is important that all consideration should be given to minimise the problems of getting around. However, most people spend only a few minutes in a car park at a time.

Figure 9.1 Disabled parking stalls

129

The design criteria for dealing with persons with disabilities within car parks are provided in BS 8300 2009. The recommendations therein, however, do not adequately distinguish between different parking functions, for example, a hospital car park may well need to increase its provision for disabled drivers, as would areas with a high percentage of elderly residents. Their location can also reduce the use by disabled drivers; that is, when constructed on a steep hillside or if ample on-street parking is available for blue badge holders in the car park's immediate vicinity. It is to be hoped that, eventually, car parks will be recognised as a separate building type in this respect and treated accordingly.

9.2. Stall locations

When endeavouring to provide stalls for disabled people, designers should be aware of the following problems.

- When located on floors other than the one that leads directly to the main exit point, wheelchair users must use the lifts. In the event of an electrical failure or mechanical breakdown, they will either be constrained to the floor on which they have parked and have to drive back out, or be unable to return to their vehicle.
- Lift doors must be wide enough to accept wheelchairs and double-width prams (800 mm is adequate for most wheelchair users but 1000 mm could be required to accommodate some prams).
- Locating disabled driver stalls at random throughout the building means that drivers must be prepared to search the entire parking area even though there may be standard-width spaces available on the lower levels. By the time that they have reached the farthest level, stalls on the lower levels may have become available and they must start all over again.
- Supervision of randomly located stalls is made more difficult and hence, their misuse by other drivers becomes easier.
- Occasionally, it is prudent to adopt a pragmatic approach and accept that a smaller number of spaces will be made available on the most popular level (say 2 to 3% of the static capacity) and have another block of 'overflow' stalls that satisfy the regulations located in a less popular location. Effective supervision will be simplified and the fewer empty stalls to be seen at the entry level will reduce the annoyance of those who are queuing to enter.
- Provided that problems of power failure, lift occupancy, extended supervision, and vehicle travel distances can be accepted, stalls may be located on any level as long as they are close to a lift core. They should be located proportionate to deck capacity, but in any case no fewer than four in a block. If possible, an indicator system that informs drivers of the parking status on other levels will improve circulation efficiency.

9.3. Stall dimensions

Stalls are, typically, 3.6 m in width with a crosshatched lane 1.2 m wide at the ends, continued along the face of the parking block. They can also be 2.4 m in width provided that a 1.2 m cross-hatched area is located between every parking stall. It should be appreciated that the stall-end cross-hatching remains part of the traffic aisle and is not additional to it. Stalls for assistants, or for parent and child, can be indicated in a similar manner to the 2.4-m-wide parking stalls for disabled drivers, but with a different legend. In this case, however, the cross-hatched end lane can be omitted (see Figures 9.2 and 9.3).

Figure 9.2 Disabled bay layout compliant with BS 8300

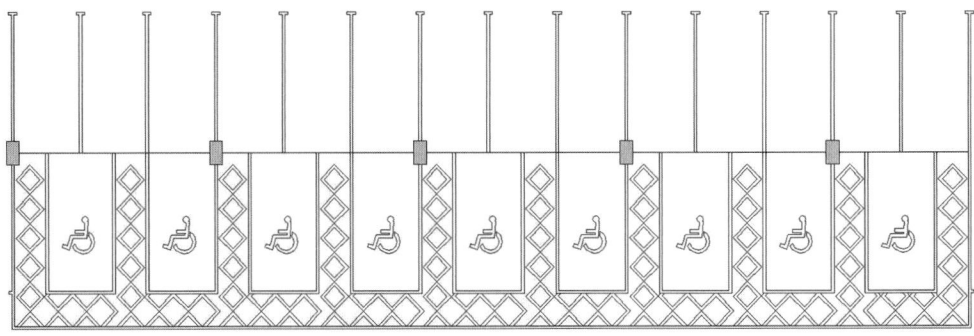

Figure 9.3 Disabled bay layout compliant with DfT and Network Rail standards

9.4. Access

In some cases, it may be prudent to separate disabled people's spaces from the main parking area by introducing a separate access and egress point. Disabled drivers, understandably, tend to reduce the inflow rate; in large car parks that are used intensively this could pose a problem. It is also of benefit for disabled drivers to readily identify the special provision, be confident of gaining access and, possibly, have direct access to a Shopmobility unit. Differential tariff options can also be introduced at the discretion of the operator.

Car Park Designers' Handbook
ISBN 978-0-7277-5814-9

ICE Publishing: All rights reserved
http://dx.doi.org/10.1680/cpdh.58149.133

Chapter 10
Bicycles and motorcycles

10.1. Discussion

The growth in motorcycle ownership and other powered two-wheelers (PTW) has created an ever-increasing pressure on the need to provide adequate facilities for their safe and secure parking. A study of the London Congestion Area found that 33% of the total capacity of 'on street' parking was occupied by motorcycles (see Figure 10.1).

Motorcycles and bicycles share the highway with four-wheeled vehicles, In an open road environment, as a result of their relative lack of manoeuvrability and protection for the rider, they are much more vulnerable to accidents than other types of road user. They are considerably narrower than four-wheeled vehicles and are able to manoeuvre in smaller areas, but they are far less stable at low speeds.

Within most parking facilities, vehicles are parked at right angles immediately at the sides of the traffic aisles and, in the main, have to reverse across the aisles in order to turn towards the exit. The relative narrowness of traffic aisles and the problems of low-speed control for two-wheeled vehicles can put them at an even greater accident risk than when on the streets outside. In the case of angled parking with even narrower aisles, the problem is exacerbated.

Over recent years an increase has been noted in proposals for bicycle and motorcycle parking within car parks. Some have been sensibly located at street level and provided with separate entry and exit points, while others have been proposed wherever there has been spare space on any or all parking levels and must share the car access. There is no legislation covering this provision and it is generally left to the designer or the client to specify the numbers and locations that they consider to be reasonable. Often, it is a case of what can be fitted in without losing car spaces.

On safety grounds alone, it is good policy to separate two- and four-wheeled vehicles by the provision of separate parking zones, each with its own access location. Motorcycle, PTW and, in particular, bicycle provision, should only be considered adjacent to the main vehicle access level. It also needs to have direct street access to be attractive to riders.

10.2. Bicycle parking

Often a requirement by planning authorities, bicycle parking needs to be designed so that its location does not interfere with pedestrian movements.

Typically, bicycle parking is free of charge, and anchor frames such as Sheffield stands are used to enable bicycles to be secured (see Figure 10.2). An alternative is to provide a lockable security cage that provides complete protection against theft and vandalism. This is particularly effective for commuter use in railway stations. They do, however, require more space for any given number of bicycles than the Sheffield stands (see Figure 10.3). Lockers for use by cyclists can also be provided (see Figure 10.1(b)).

10.3. Motorcycle parking

Motorcycles are self-supporting and can be restrained at the rear (see Figure 10.4). However, they are difficult to manoeuvre when being manhandled and consideration should be given to pushing them forward into and out of the stand. Large motorcycles occupy a space of about 2800 mm × 1300 mm. Side by side they can be located at a spacing of 1300 mm. PTWs of smaller size can, if necessary, have their own reduced dimension parking area with a bay size of 2500 mm × 1000 mm. However, supervising

133

Figure 10.1 (a) Bicycle and (b) motorcycle parking

(a)

(b)

the use of different-sized parking stands can be an unrealistic exercise. It is recommended that the areas be marked as a zone, rather than bays, unless it is intended that they be used as metered bays.

The resting surface should be hard and sufficiently strong so as to resist the point load exerted by the motorcycle stand. Standard tarmac exposed to solar gain can soften in hot weather and materials such as grass blocks should be avoided. The surface should have good drainage with gullies kept clear of the bays to avoid loss of dropped keys and be near level to avoid motorcycle instability. The site of bays should be close to the main

Bicycles and motorcycles

Figure 10.2 Anchor stands enable bicycles to be secure

Figure 10.3 Dimensions of 'Sheffield' stand

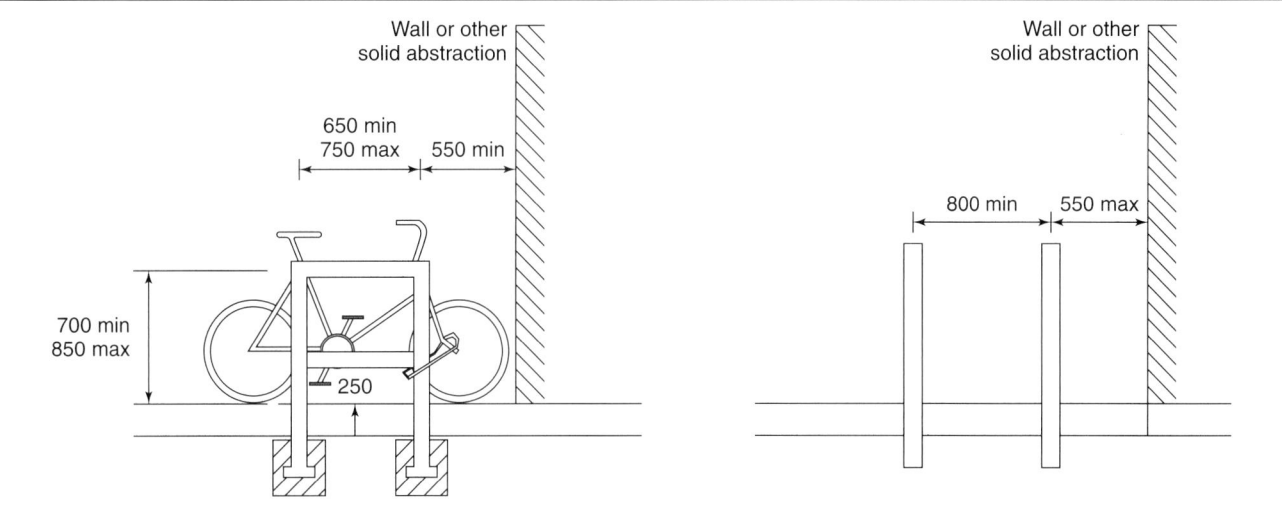

Figure 10.4 Motorcycles can be restrained at the rear

entry point to attract use and be in an area subject to casual surveillance by other bikers and passers-by. To add to the security, adequate lighting and inclusion on CCTV coverage should be included.

10.4. Lockers

A major problem for many motorcyclists and some cyclists is finding somewhere for their special clothing, helmets, leathers, boots, rainproof wear and so on. Metered anchors with helmet lockers are an option as are lockers situated close to supervised areas to reduce the incidence of theft and vandalism.

10.5. Fiscal control

It is unreasonable for motorcyclists to have to pay the same amount for their parking as motorists and yet, if they were to proceed through the vehicle control barriers in a Payment on Foot system, they would incur the same charges. A separate entry and exit allows a differing charge rate. A flat payment could be charged either on entry, or exit, or by a Pay and Display system.

Car Park Designers' Handbook
ISBN 978-0-7277-5814-9

ICE Publishing: All rights reserved
http://dx.doi.org/10.1680/cpdh.58149.137

Chapter 11
Security

11.1. Discussion Security or the lack of it within some car parks is a major issue with the parking public. The statistics indicate, however, that personal danger is more perceived than actual. Feelings of insecurity that can be engendered when walking through a dimly lit car park late at night are not dissimilar to those developed when walking through, say, a dark street on a dull evening. The actual danger is probably minimal, but the perceived danger creates fears that, if not allayed, could result in the car park being shunned by many motorists and in particular women (see Figure 11.1).

The use of CCTV is now commonplace and has led to a great leap forward in car park security. However, it cannot see around bends or corners. Curvi-linear or circular aisles are restricted in this respect or require more cameras, thus more curves mean more cameras, more screens, more staff and greater expense (see Figure 11.2).

Feelings of security are also enhanced by a reduction of internal vertical structure. Structures spanning clear over each bin are preferable to those where columns and shear walls are located adjacent to the aisles, behind which potential felons can lurk. It has been demonstrated, by competitive construction over the years, that properly designed clear-span structures are no more expensive to construct than those with internal columns.

The Park Mark scheme promoted throughout the UK provides an award system for approved facilities with specialist advice from police and other security experts. It stipulates a basic standard and, when the Park Mark is displayed, it may confer confidence on car park users.

Figure 11.1 'Help Points' provided on each deck

137

Figure 11.2 CCTV camera

11.2. Lighting, music and CCTV

The requirement for general lighting is covered in Chapter 13. Good lighting enhances the feeling of security, whereas dark areas do not and should thus be avoided. Soft relaxing music can be helpful, especially late at night, and the knowledge that CCTV cameras are supervising one's movements can be very reassuring. A music delivery system implies a public broadcast capability and thus a human presence. It also helps if two-way communication can be achieved between the customer and the supervising staff by way of 'Help Points'. Such measures can go a long way towards dispelling feelings of insecurity for both men and women when they are in these buildings, and can pre-empt moves to segregate the sexes in different car parks.

A major advance in the improvement of car park security has been the adoption of CCTV. Advances in camera technology have led to better resolution and less light sensitivity: with improved surveillance from a fixed vantage point it is easier to record antisocial activities and respond appropriately to risky situations. To make the most efficient use of this high-tech surveillance, the car park needs to be designed appropriately. The presence of surveillance cameras should be highlighted with signs and visible cameras. An efficient layout should eliminate blind corners and reduce the incidence of internal obstructions such as columns and walls to an absolute minimum, especially when they are next to traffic aisles. Overly short distances that make inefficient use of the system should be avoided. Overcoming design inadequacies merely by installing more cameras does not readily solve the situation. It has been established that the maximum time that operatives can monitor up to six screens effectively is about two hours. After that, they need to step down for two hours at least. The more screens that have to be monitored, the more rapidly is an operative's effectiveness reduced, to the extent that a display of antisocial behaviour may not be recognised as such. Increasing the number of surveillance staff might solve the problem, but the car park running costs would also increase.

Circular car park layouts are the most inefficient shape to monitor, followed by curvilinear forms which, although they may be visually attractive, complicate security issues. Split-level layouts restrict the width available for surveillance; the best shapes for security are those with straight decks and a construction system that spans clear over traffic aisles and their adjacent stalls (generally, 15.6 m for one-way flow layouts and 16.5 m for two-way flow layouts.)

Security

Figure 11.3 Glazed wall/doors to lift/stair lobby

For security to be really effective it needs manpower, to provide a visual presence, undertake surveillance and be on call to assist; informative signing also helps. Knowing precisely where to go for assistance can relieve a sense of unease. (Signing issues are covered in Chapter 14.)

11.3. See and be seen

One of the major concerns is that people might enter an enclosed stairwell and be assaulted without being seen by others. CCTV can play a major role in forestalling this, but the provision of glazed areas that enable any potential attacker to be exposed to public gaze also helps. Wherever there are areas where the public is vulnerable to attack, glazed doors, windows or glass walls are desirable. There should be no hiding place for antisocial behaviour. Vision panels in lifts and lobby doors onto decks should be provided, and at a height that patrons in wheelchairs can see through (see Figure 11.3).

11.4. Full bay surveillance

Ever more sophisticated systems are being introduced to improve car parks and at the higher end of the level of service are systems with display vacant or occupied parking bays and then monitor, on camera, that bay with ANPR (Automatic Number Plate Recognition). This enables individual cars to be automatically registered and movement, authorised or not, to be recorded. It also allows a motorist who has forgotten where they parked to use a Help Point display; by entering all or part of the number plate the computerised system will display the floor and bay where the vehicle is, or indicate its departure. Systems of this nature can therefore give individual bay surveillance without the need to increase general staff numbers (see Figure 11.4).

11.5. Ground-level enclosure

It is advisable to control entry and exit points for a stand-alone car park to allow surveillance of pedestrian traffic, and to monitor all persons in the event of an incident.

Figure 11.4 Parking bay indicators use a green or red light to indicate whether vacant or occupied

The car park sides should be enclosed with a security fence while maintaining fire escape provision requiring all legitimate car park users to enter by way of designated points. This also has the benefit or preventing random walk-through routes being used in what is the most heavily trafficked parking level.

11.6. Women-only car parks

One apparent solution to feelings of insecurity among women when using car parks has been the introduction of women-only car parks.

Put forward largely by politicians, the subject is aired publicly every few years, but such a proposition would be impractical for a number of reasons.

- What is to prevent a man from cross-dressing and driving into the car park? What socially acceptable procedures could be implemented to prevent this happening?
- Most car parks are not difficult to enter by way of the open sides or fire escape doors, or even hidden in someone's vehicle.
- The operator will become even more responsible for the safety of the user. Construction, supervision and insurance costs will rise, to be reflected in increased parking charges.
- Who, or what authority, is prepared to underwrite the cost of introducing such a scheme?
- Will there be enough women drivers of a nervous disposition in any one place to justify their introduction?
- The location of such a car park advertises where women can be found alone in the streets outside as they make their way to and from it. It will become a focal point of attraction for those who prey on the female sex.

In the opinion of the authors, proposals for women-only car parks should be resisted for these reasons.

Car Park Designers' Handbook
ISBN 978-0-7277-5814-9

ICE Publishing: All rights reserved
http://dx.doi.org/10.1680/cpdh.58149.141

Chapter 12
Underground and robotic parking

12.1. Discussion

The tenets contained in Sections 3 to 9 are the same for underground car parks as for those above ground. In practice, the only difference is in the design of the sub-structure. In all other respects they can be considered to be the same as for an enclosed car park above ground. The main problems that occur are to do with earth pressure and keeping water out. The provision of mechanical ventilation and fire and smoke control is given in Chapter 18. Unlike open-sided car parks, fire spread from vehicle to vehicle is of major concern when considering the fire load in an enclosed underground car park and as such an increase in the standard of fire protection to the structure will be required (see Figure 12.1).

The drum shape of a circular car park is well suited to being located underground and the unused space in the middle can provide ventilation, services and escape facilities. However, such layouts cannot, practically, incorporate separated flow routes and so, their capacity should be limited to 400 spaces when used by the general public. When large-capacity layouts are proposed the layouts will not differ greatly from those proposed above ground.

The cost of an underground car park per car space, in general terms, is at least twice that for an efficiently designed, open-sided, car park above ground but in difficult ground it will be even more. Running costs are also higher due to the increased use of fans for mechanical ventilation.

In many countries, underground car parks are state subsidised but in the UK public car parks usually have to be paid for by commercial enterprises or the general public. Other

Figure 12.1 Entrance to underground car park in Manchester

than in exceptional circumstances, such as the centre of a major conurbation where the need is great, land is at a premium, and a high user charge is achievable, it is difficult to justify their construction on economic grounds, no matter how desirable they might be on environmental grounds.

12.1.1 Automated car parks

Automated car parks were first developed in the United States in the early 1920s but robotic car parks took off in a big way in Japan during the 1970s.

Robotic car parks are ideally suited for congested city centre sites where the following factors can be considered advantages (see Figure 12.2).

- Inadequate space for a conventional multi-storey car park.
- High land cost/maximise value.
- High parking demand.
- Ability to 'cloak' for appearance.
- Avoids public areas to the superstructure.
- Suits non-peak loading.
- Efficient 'no contact' parking.

Figure 12.2 Internal views of a robotic car park

- Low emissions from vehicles.
- Does not need open sides for ventilation.

Disadvantages are

- higher cost to construct
- high cost to maintain
- restricted access and egress delivery and recovery times
- requires a manned site at all times
- less convenient for some users, for example, when returning to offload shopping, or meeting children's needs
- needs entry/exit plaza for loading and retrieval.

These will need to be procured by specialist suppliers.

A typical layout for a robotic car park is shown in Figure 12.3.

Where small sites are being considered for development, especially in the case of private parking associated with apartment blocks, the use of drive-through vehicle lifts can be used to gain access between parking levels. A typical layout for such a lift is shown in Figure 12.4.

Figure 12.3 Typical layout for a robotic car park

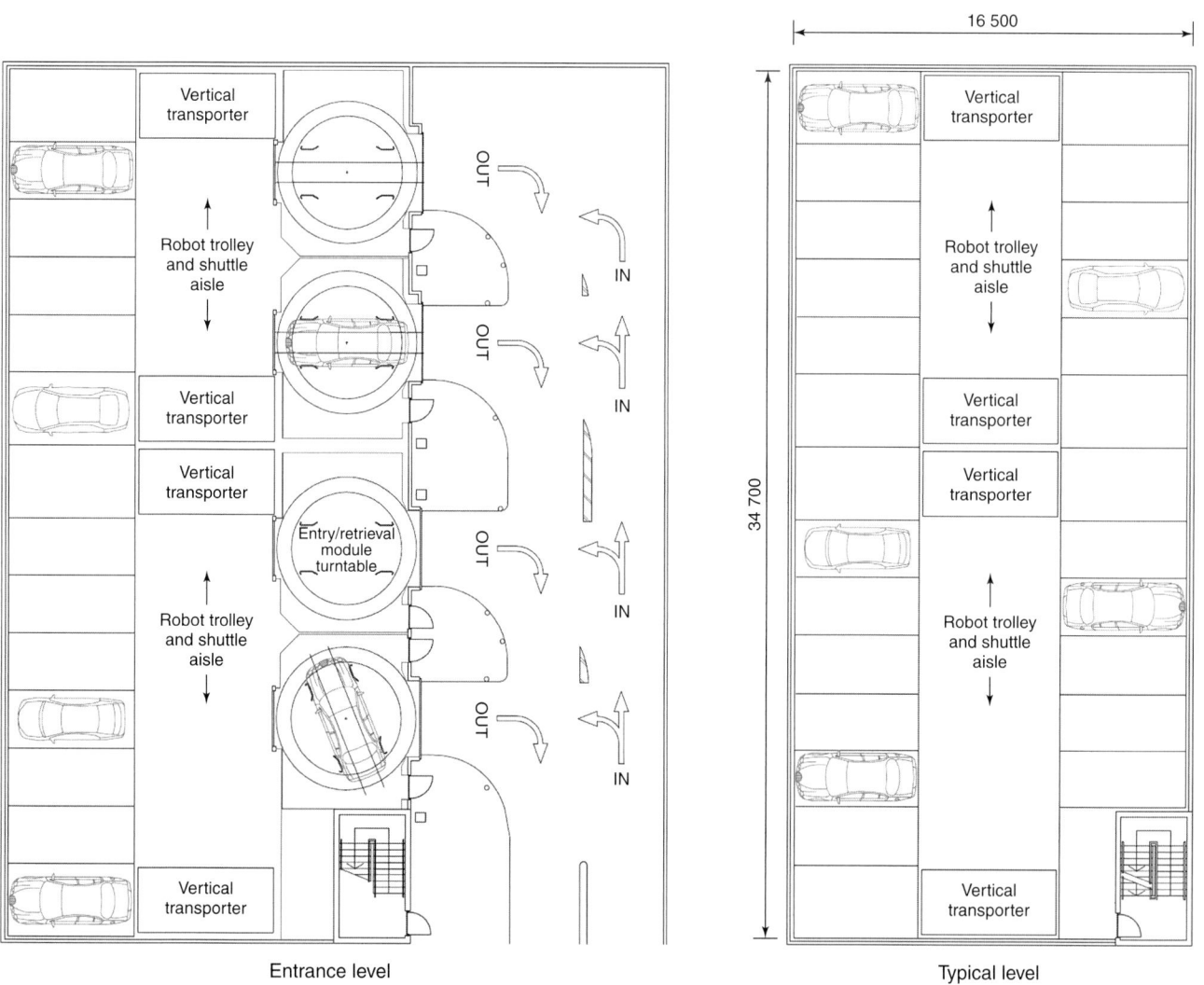

Figure 12.4 Typical dimensions for a single car lift

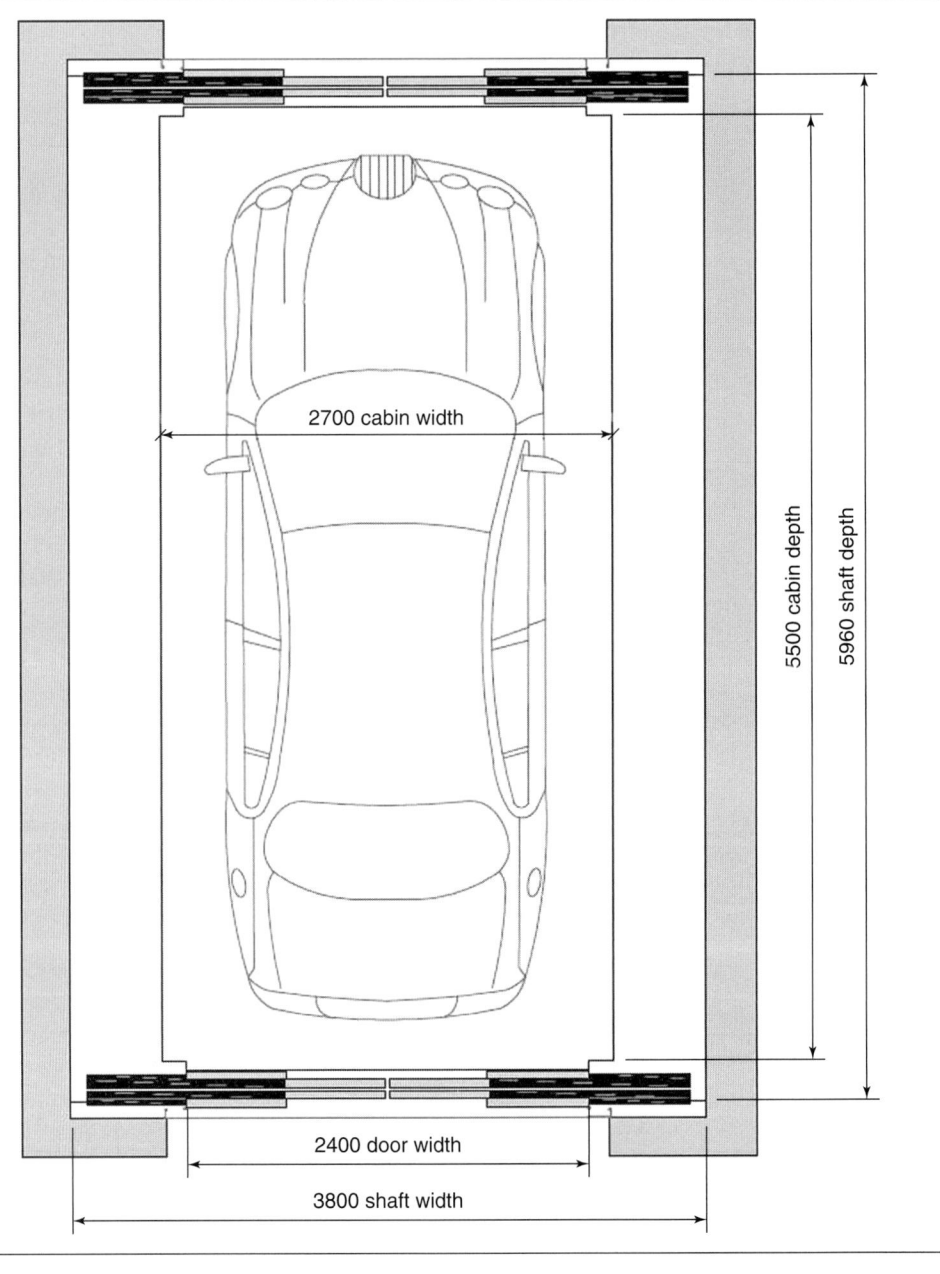

Car Park Designers' Handbook
ISBN 978-0-7277-5814-9

ICE Publishing: All rights reserved
http://dx.doi.org/10.1680/cpdh.58149.145

Chapter 13
Lighting

13.1. Discussion

The commercial viability of a car park can be affected by the initial impression of users, and one way to create a positive impact is to have good uniform lighting which helps to create a positive effect. There is guidance to minimum lux levels to be used by the designer, backed by the Park Mark scheme (see Figure 13.1).

Consideration should be given to luminaire degradation over time and to schedule for periodic renewal to maintain a minimum level of lighting. Energy use can be reduced by use of suitable controls or, indeed, choice of fitments. Accounting for daylight levels and measures to monitor usage with proximity sensors and so on will affect the overall energy efficiency. Modern developments in LED lighting systems have dramatically reduced power consumption. The low headroom levels of most car parks will present a challenge to the maintenance of uniformity, and reference to the structure will determine the number of fitments to achieve the required uniformity (see Table 13.1).

Reflective surfaces such as painted soffits will give an uplift and will form part of the design considerations. Luminaires themselves should be compact so as not to affect headroom and to be tamper proof. And good practice is to minimise the extent of exposed service conduits.

A car park is predominantly an open-sided structure and night light pollution may impact on the local environment; the design should seek to minimise light spillage, which will also

Figure 13.1 Luton Station multi-storey car park at night

Table 13.1 Recommended luminance for multi-storey and underground car parks

Area	E_{ave} (lux)	E_{min} (lux)
Parking bays, access lanes	75	50
Ramps, corners, intersections	300	120
Entrance/Exit zones (vehicular)	Night: 75 Day: 300	See text
Pedestrian area, stairs, lifts	100	50
Disabled parking	300	n/a
Roof level parking in Rural Zones E1 and E2	15	6
Roof level parking in Urban Zones E3 and E4	30	12

NB: Uniformity E (min) to E (ave) greater than 0.4 as in BS EN 13201[6.4]

have an effect on efficient energy use. The top and exposed deck will have the lighting set to a lesser lux level than contained floors, and can also use pole-mounted directional lighting to restrict overall light spillage. The height of lighting masts may well be subject to planning requirements.

13.2. Zoning

The entrance and exits to a car park present the motorist with a transition from exterior to interior, both by day and by night, and a change in levels and appropriate highlighting level is required to cater for the transition. This will also be the point where drivers stop and start their vehicles and need to deal with barriers, ticket machines, signage and other operations related to the car park.

Vehicle ramps and main transition routes present the next area, requiring a good light level to emphasise changes in road surface, kerb edges and manoeuvring positions. Where there is a possibility of encountering pedestrians the designer needs to make sure that there is adequate illumination and, in particular, uniformity of lighting levels, to minimise shadow and show up persons in dark clothing so that a contrast is achieved.

Common areas need sufficient lighting to make the parking manoeuvre easy and for exiting of a vehicle. For security, adequate lighting is necessary to enable CCTV cameras to operate and to avoid dark areas where there is a potential for criminals or vandals to lurk and create problems.

Pedestrian areas within the car park should be subject to light levels that clearly give direction to access cores; within those transit areas, to allow people to leave and return to the car park, there should be levels enabling them to clearly see and be seen. The various levels of minimum luminescence will satisfy the basic design criteria and enable the installed system to perform to acceptable levels.

13.3. Emergency lighting

Illuminated signs should be installed at locations such that in the event of a power failure at night they can guide pedestrians to the nearest escape route. They should incorporate a backup battery power located either in each individual sign or from a central battery or emergency generator source.

Emergency lighting should conform to the requirements of BS 5266: Part 1 (2011b).

Car Park Designers' Handbook
ISBN 978-0-7277-5814-9

ICE Publishing: All rights reserved
http://dx.doi.org/10.1680/cpdh.58149.147

Chapter 14
Signage

14.1. Discussion

Signs and notices should advise, direct, and be easily understood and easily recognisable. Even when travelling at low speeds in an enclosed area the driver's workload can be relatively high and a proliferation of difficult-to-read signs and notices only serves to confuse. Conversely, too few directional signs can create a sense of unease. The best circulation layouts are those where the need for signage is minimal.

To cater for drivers with little knowledge of English, signs should, where possible, be graphic and incorporate symbols readily found on the public highway. One-of-a-kind graphics should be reserved for commercial branding or other site-specific uses. Guidelines should be consulted for pedestrians with visual impairment, concerning measures such as the use of Braille and large print, and for disabled drivers, such as the positioning of lift buttons, viewing panels and so on. There will also be a need to provide signs for protection of liability, for example, instructions for people not to leave valuables in their vehicles, or fill them up with petrol, and such signs should be located so as not to conflict with other direction or advice signs.

To assist the motorist there should be a combination of deck markings and overhead signs located at key changes of direction. In one-way-flow layouts, signs over the traffic aisles showing single arrows in the direction of the traffic flow and 'No Entry' symbols which meet drivers going in the wrong direction, should occur frequently. Pedestrians need to identify the level they are on and then be directed towards the principal access core. This can be helped by colour-coding each floor level and having floor-level indicator signs immediately outside the lift doors. If there is more than one access core then each should be identified with a unique reference such as the street name onto which it exits. Working on the basis that some motorists will be unfamiliar with the car park, clear directions to the inflow search path will be required, with advice as to whether the traffic aisles are one-way or two-way and, preferably, indications of the parking status in traffic aisles adjacent to the main circulation route (Variable Message Signs). This encourages users to bypass full areas and head for the vacant spaces available further into the facility.

14.2. Directional signs

Direction arrows are a familiar symbol for deck markings in conjunction with overhead signs. Arrows painted on the decks are viewed obliquely and need to be large and long. Most turns will be at right angles and the arrow sign showing this should be clear and unequivocal. Where a large car park incorporates a rapid inflow route that can bypass full or traffic-congested decks, variable message signs will also be useful in advising motorists to avoid the aisle ahead and take the rapid route to another deck level. This will improve dynamic efficiency and assist in reducing driver frustration.

Motorists will also require directional arrows coupled with 'EXIT' signs when leaving the car park. Where a rapid and 'excluded' outflow route is incorporated it should be clearly marked at each direction change to avoid mixing with the inflow route (see Figure 14.1).

Pedestrians readily understand signs incorporating a walking man, coupled with an arrow showing the direction of travel. It is also common practice to provide a 1 m-wide painted strip on the deck with the walking man graphic to one side of the traffic aisle and leading to the access core. Pedestrians are made to feel secure by this action, but it

Figure 14.1 Signs at a station car park

should be borne in mind that the painted strip remains part of the traffic aisle and they are being channelled immediately behind the parking bays. It renders them vulnerable to vehicles leaving their parking bays, particularly when vehicles are reversing. Children running ahead or lagging behind their carers can be obscured from the driver's view and are put at risk (see Figure 14.2).

Figure 14.2 Dedicated pedestrian walkway

With right-angled parking and a one-way traffic flow, pedestrians and cars share a 6 m-wide aisle and have a reasonable aisle-width in which to see and be seen. When angled parking is adopted, however, aisle widths, for, say, 45° parking can reduce to 3.6 m, which reduction places vehicles and pedestrians in close proximity. The 1 m strip with a 'walking man' symbol cannot easily be used with angled parking arrangements.

Control signs such as 'No entry', 'Permit parking only' and 'Authorised disabled drivers only', and other, similar signs should be designed such that they stand out clearly from the directional signs. Where signs have highway-approved equivalents showing the maximum allowable speed – 'Give Way', 'Stop', 'No entry', and so on, the highway designs should be adopted so that drivers will more readily conform to them.

14.3. Information signs

Motorists and pedestrians need to know precisely where they are parked. When in a hurry to keep an appointment, catch a train, or simply to deal with children, it is easy to forget to note the whereabouts of a particular car on a large-capacity deck, and even the floor level on which it is parked. Numbers painted at the ends of bays assist in locating the vehicle on a particular floor and in really large-capacity layouts, block zoning (A, B, C, etc.) can also help. The colour coding of individual levels also helps drivers to remember their parking floor level.

Overhead signs specifying the deck level and, where applicable, the specific zone, provide a good reminder and themes showing unusual characters, such as bears, giraffes and horses or, boats, trains and planes, can be stencilled around a parking floor. If they are repeated sufficiently frequently they can be a good 'memory jog'. The repeated use of colour as bands on columns and walls also aids recall. Signs showing the parking level together with the appropriate character should be prominently displayed at the entrances to the access core, with repeat signs within the core.

The location of pay stations in Payment-on-Foot and Pay-and-Display systems needs to be shown, preferably by the use of overhead signs that can be seen from a reasonable distance. With a Payment-on-Foot system drivers need to be informed that they need to pay before returning to their vehicle. The preferable location for this is at the entry point and also when they first enter the stair core as pedestrians. A plan of each floor-level is useful in larger car parks to provide information on destination points.

Advice signs to conform with legislation or limit the operators' liability should be located so as to be visible, but neither conflict with key directional signs nor distract motorists. The type of sign and colour should not be the same as the direction and principal information signs (see Figure 14.3).

Technological advances have assisted way finding and integrated CCTV surveillance, and user help systems can register vehicle number plates and their location within the car park. With large car parks, especially in shopping malls, disorientation can occur and in order to locate one's car, part of the vehicle's registration is entered in a Help Station, whereupon the system identifies the car and its location and shows a suggested route. The system also comes with a light display to indicate green for a vacant stall and red for occupied, with options for blue for disabled drivers' bays and orange for bays for parties with children.

14.4. Variable message sign systems

These consist of metal sensitive loops or infrared beams built into ramps, access-ways or stalls. Connected to electronic signs, they provide information on the parking availability for a complete facility, deck, section, or even an individual stall, depending on the chosen system. Vehicles, passing over these loops or through beams, activate the relevant signs, informing drivers of the parking status and preventing needless circulation through areas that are already full with parked vehicles. Their inclusion can often be of significant benefit to circulation efficiency.

14.5. Emergency signs

Fire escape and emergency exit signage must conform to BS 5266: Part 1 (BSI, 2011b) and be located in appropriate positions, together with additional directional signs. They must

Figure 14.3 Typical signage schedule

Ref	Sign	Colour	Size
1	 FLOOR LEVEL 4	Legend and borders white Background to BS 4800, colour coded to suit	300 mm to 500 mm high
2	 Fire door Keep shut	Blue on white, self-adhesive	Proprietary item
3	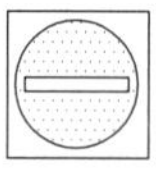 (no entry symbol)	Legend and borders – white Background – red	300 mm²
4	 ← ONE WAY	Legend and borders – white Background – blue	600 mm × 300 mm
5	 (no right turn symbol)	Border and diagonal – red Background – white Legend – black	300 mm dia.
6	 🚗 EXIT →	Legend and symbol – black Background – white Border – blue	1450 mm × 250 mm
7	 P → Note! Arrows can point in any direction	Symbol – black Background – white Border – blue White P on blue background	600 mm × 250 mm
8	 Variable message	Symbol – black Background – white Border – blue	900 mm × 300 mm
9	 PAY STATION AT LEVEL ?	Legend – black Background – white Border – blue	400 mm × 300 mm
10	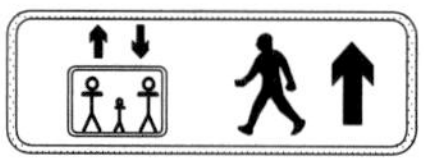 (lift and pedestrian symbols)	Legend – black Background – white Border – blue	1450 mm × 300 mm

be illuminated and incorporate rechargeable batteries that, in the event of a power failure at night, will enable pedestrians to find their way to a fire escape.

Car Park Designers' Handbook
ISBN 978-0-7277-5814-9

ICE Publishing: All rights reserved
http://dx.doi.org/10.1680/cpdh.58149.151

Chapter 15
Drainage

Drain locations should be considered during the design process and not added in as an afterthought. They should be inconspicuous, easy to maintain and located where they cannot suffer damage from vehicles (see Figure 15.1). When located where damage might occur, protective measures such as hoops or protective posts should not be located such that they intrude into the parking stalls, especially where they reduce the available width.

In the UK, drainage should be designed in accordance with BS EN 752 and BS EN 12056 (BSI, 2000)). Exposed roof parking decks located directly over shops, offices or other habitable areas should be designed for a full storm intensity (75 mm per hour) to ensure the system is capable of coping with a flash flood, otherwise a lower intensity can be

Figure 15.1 Protected drainage

Figure 15.2 Attenuation tank under construction

used (50 mm per hour). Stair cores need special attention to ensure that drainage falls do not channel water into them, particularly as mobility regulations have negated the use of thresholds.

It should not be necessary to provide U-bends on outlets from the upper parking decks. It should be sufficient to provide this feature on the lowest level only. Open-sided car parks have their gully outlets vented to fresh air, even when they are within the perimeter of the building. On the intermediate decks water occurs only intermittently allowing material on the underside of vehicles to drop off, be washed towards the drains where, if U-traps are incorporated, it solidifies, builds up and can, eventually, block them. Occasional maintenance, if carried out, can overcome the problem, but U-traps are an unnecessary feature.

Some authoriities' regulations require that the roof deck drainage be separated from that on the intermediate decks. The incidence of petrol and oil leaks is low at any time but it is argued that such leaks into the drains from intermediate decks can be more concentrated than from a weather-exposed roof deck, even though extended periods of dry weather can occur. A petrol interceptor is required to intercept the intermediate located drains before the system enters the main sewer, but if the roof drainage is to be separated it means that twice as many down-pipes will be necessary.

In the interests of holding back storm water run-off from entering water courses, most new developments now require stormwater attenuation (see Figure 15.2).

The large surface area of a ground slab to a multi-storey car park is ideal for the location of attenuation systems which cater for storm water-run-off in the short term, during a storm. Through-flow control valves allow the discharge of storm water into the drainage system in a controlled manner.

Minimum drainage falls of 1.67% (one in 60) are recommended on exposed decks, although 2% (one in 50) are to be preferred to better contend with structural deflections. Where joints occur between decks and are used for movement it is sometimes possible to leave them open and provide a drain underneath, to catch the rainwater. Proprietary cast-in-slot drains are, visually, less intrusive, but can be difficult to make watertight, especially if used in suspended slabs. On some awkwardly shaped sites it is desirable to rotate part of the structure through 90° or more. Unless carefully considered, this could lead to problems in creating falls that are compatible at the change of direction.

Facilities for draining and washing down decks should be provided at each level, and spaced such that the length of hosepipe to be used can cover all of the deck area.

Car Park Designers' Handbook
ISBN 978-0-7277-5814-9

ICE Publishing: All rights reserved
http://dx.doi.org/10.1680/cpdh.58149.155

Chapter 16
Fire escapes, safety and fire fighting

16.1. Discussion

Car park fires by their nature are relatively low-risk and well defined, and their occurrence is relatively rare. The following table is reproduced from the BRE Building Regulations Division Project Report *Effects of Local Acts on Fire Risks*. It shows the statistics of car park fires in the UK, for places that have Local Act fire safety provisions in place, and for the one place that does not (Blackpool) (see Figure 16.1).

The report covers statistical analysis based on UK Home Office Fire Statistics databases 1994–1999.

The statistical data supplied in Table 16.1 serves to highlight the relatively low occurrence of fires within car parks. For regions without Local Act fire safety provisions, the

Figure 16.1 Fire fighting

Table 16.1 Results of statistical analysis of car park fires

Results of car park fires	Local Acts (number of people) (per million population)	Number of acts (number of people) (per million population)
Deaths	0	0
Injuries	0	0.1
Fires >5 m²	0.3	0.8

Source: BRE Building Regulation Division Project Report *Effects of Local Acts on Fire Risks*

results show that per million population, an average of only 0.8 fires that occurred were greater than $5\,m^2$ in area. Additionally, there were no recorded deaths during the five-year period.

There is also very little evidence to suggest that car park fires in well ventilated open-sided car parks typically spread to involve an entire floor plate. This is recognised in other UK fire safety guidance such as Approved Document B (NBS, 2010b), in which clause 11.2 suggests

> Buildings or parts of other buildings used as parking for car and other light vehicles are unlike other buildings in certain respects which merit some departures from the usual measures to restrict fire spread within buildings.
>
> Those are:
>
> - The fire load is well defined
>   And;
> - Where the car park is well ventilated there is low probability of fire spread from one storey to another.

NBS, 2010b

This data suggests that due to the low occurrence of large fires within open-sided car parks the risk to life safety due to fire in open-sided car parks is low.

16.2. Means of escape

The basic principle for the design of means of escape are that there should be alternative means of escape so that users of the car park can turn their backs on fire, smoke, or fumes, and escape within a prescribed travel distance to a place of safety.

A place of safety in a multi-storey car park would typically be an enclosed escape stair of which a minimum of two will be required. Note in the case of a split-level car park, alternative escape routes will be required from the decks either side of the split level unless DDA-compliant pedestrian routes are provided across the split level. The standard vehicle ramps used in split-level design are not suitable for use as pedestrian escape routes.

Maximum permitted travel distances for pedestrians can be found in Table 2 of the Building Regulations Approved Document B1 (NBS, 2010) where car parks are classed under Purpose Group 7 (storage and other non-residential) – normal hazard. This refers to maximum travel distances of 25 m for escape in one direction and 45 m where escape is available in more than one direction.

On top decks where escape is in the open air then these distances can be increased to 60 m for one direction and 100 m for escape in more than one direction.

Using the recommendations given in Table 12 of BS 9999::2008 (BSI, 2008a) and for a Risk Profile of A2, the one-way travel distance for intermediate decks would be 22 m but for two-way travel this would increase to 55 m.

16.3. Fire safety

Fire engineering predictions and an assessment of the likely conditions that will occur during actual fires should be carried out, together with an ongoing fire safety management system: a particularly relevant matter when the car park adjoins another building use such as a shopping centre, leisure centre, or cinema. A typical fire safety strategy for a car park scheme should incorporate the following elements.

- A suitable fire alarm system using manual-break glass contacts at each fire exit with auto-detection in lift shafts and lift motor rooms.
- A means of control of smoke and fuel vapour hazards, usually combined with the proposed method (natural or mechanical) of ventilating the car park.
- Means of escape calculations for exit provisions (see Figure 8.2).

- Escape lighting (in accordance with BS 5266-1:2011 Part 1 (BSI, 2011b)) to provide luminaries continuously energised for up to three hours by internal or a central battery standby, to give full or reduced lighting output from the lamps of the luminaries.
- Maintained exit and emergency fittings provided over door exits to stairwells, changes in direction and fire exit doors.
- Provisions for disabled pedestrians: escape routes suitable for their use, and refuges in an enclosure providing resistance to fire for up to half an hour (on escape stairway landings).
- Compartmentation between car park decks and fire escape stairs, lift shafts, basement floors, and adjoining buildings containing different fire fighting measures.

16.4. Fire-fighting measures

Vandalism has led to the prevention of portable fire fighting equipment such as extinguishers and hose reels from being freely available on parking floors. They are now, quite frequently, located in an attendant's office, if there is one, or in manual break-glass cabinets next to the staircase exit doors.

It has long been recognised that the fire load in car parks is not particularly high, and vehicle fires do not spread. Many fire officers nowadays ensure that the occupants have been evacuated, and then leave the fire to burn out, rather than risk the safety of their personnel, unnecessarily, in fighting the fire.

Dry-risers should be located in the escape stair cores with an external inlet box provided at an appropriate level for fire service vehicles and outlet valves for hose connection at each floor level.

When the building is more than 18 m above, or 10 m below, the fire service vehicle access level, it should be provided with fire-fighting lifts. Some fire officers have interpreted this *Building Regulation* requirement as 18 m measured to the highest enclosed or covered-over deck level, open-deck roof parking being excluded from this assessment.

16.5. Sprinklers

NBS (2010) acknowledges that it is not essential to install sprinklers. Modern mechanical ventilation systems are available that use impulse and jet fans to ensure that smoke will be contained, channelled through an air corridor and guided towards the extract point. It is now possible to keep one side of a burning vehicle clear of smoke, thereby aiding visibility and the approach of the fire service personnel.

16.6. Fire escapes

The requirements for fire escapes are provided in Figures 8.2 and 8.3.

Car Park Designers' Handbook
ISBN 978-0-7277-5814-9

ICE Publishing: All rights reserved
http://dx.doi.org/10.1680/cpdh.58149.159

Chapter 17
Fiscal and barrier control

17.1. Discussion

Parking free of charge in a public, structured car park is, very nearly, a thing of the past. Ignoring the site value, the average cost, per car space, for an efficiently designed new-build facility is £8500 to £9500 at 2012 prices. Add to that the running and staffing costs, the rates, utility charges, interest charges, an allowance for maintenance, and a reasonable profit margin, and it can be seen that £1 per hour is not an unreasonable amount to charge. Even so, that rate implies that every parking space will be used efficiently during the course of a week.

17.2. Control systems

There are several type of control systems (see Figure 17.1).

- Payment on exit – where a time-stamped ticket is dispensed on entry and handed back to an operative in a kiosk when leaving. Payment is then assessed on the time difference. The main ongoing cost of this system is that of having an operative manning the exit station at all open times.
- Pay and display – a system whereby motorists assess the amount of time that they will be staying and pre-pay accordingly. The ticket can then be affixed to the vehicle window. Fines are imposed for an over-stay and money is not returned for an under-stay. When adopted by a private operator the imposition of fines can be difficult. The ticket dispensers are relatively economical to install and barriers at the exit are not an essential feature. However, an entry barrier that prevents the car park from becoming over-crowded is desirable to eliminate the possibilities of vehicle congestion.

Figure 17.1 Barrier control

159

- Payment on foot – a similar system to pay on exit but the timed ticket is inserted into an automatic pay station when leaving: on payment of the correct amount, the ticket is stamped and can then be inserted into a ticket acceptor at the exit that operates the barrier. The ticket is time-limited to prevent motorists from short-circuiting the system. This is the most expensive type of control but is essentially fairer than Pay and Display and eliminates the possibilities of fraud where operatives are involved. A typical installation for a 750-space multi-storey car park would be as follows
    - four pay stations – a central computer with mouse, keyboard, screen, printer and UPS (uninterruptible power supply)
    - intercom facility – manually operated automatic till
    - two entry ticket spitters
    - two exit ticket readers
    - four rising-arm barriers
    - cables and loops.

The total cost of such a system will be $c.$ £80 000 (2012 prices) (see Figure 17.2).

- Payment by mobile phone – an increasingly popular method, which operates on the basis of paying for parking charges over the telephone by a call text or app. A number is given at the meter or ticket machine and a text message or phone call is made to purchase an amount of parking time. The vehicle registration number is given and details are entered into a central computer to which parking wardens can refer and check that payment has been made. Drivers can set up an account and charges are added to their mobile phone bill. Payment can be debited against a credit or debit card. Some systems are also set up to ensure that drivers receive a timely reminder that the time booked is about to expire and are given the option of purchasing more time.

Figure 17.2 Typical Pay Station

- Tag system – used extensively on toll roads in Europe and now extending into car park use. An electronic tag is inserted into the windscreen of the vehicle that can be read from cameras. It records the time of entry and exit and bills are rendered accordingly. Drivers are not required to take tickets on entry or when leaving; recognition of the tag opens the barriers automatically. It is not in general use in the UK as yet, but is gaining interest.

However, some provision would need to be made for those who wished to enter the car park but were not contributors to the Tag system (visitors from outside the UK, rented or borrowed cars, occasional drivers, etc.)

- ANPR (Automatic Number Plate Recognition): similar to the tag system but reads the number plate and assigns the driver to a payment system. A sophisticated ANPR system can be used without barriers, as vehicles are recorded in and out of the car park and where valid payment is not registered follow-up tariff collection is possible by way of the DVLC database on vehicle ownership. It is now possible to achieve high positive-recognition figures, but some plates avoid being read and thus some avoidance has to be accounted for.

17.3. Barrier control

Rising-arm barriers are provided to performance criteria as specified in BS 6571 (BSI, 1989). The rising arm should have a fracture plate or be mounted on a clutch system such that it can be broken off or pushed up in an emergency. The lower edge of the arm should be cushioned so as to minimise damage if the arm descends on a pedestrian.

Figure 17.3 A typical barrier layout incorporating a central bi-directional lane

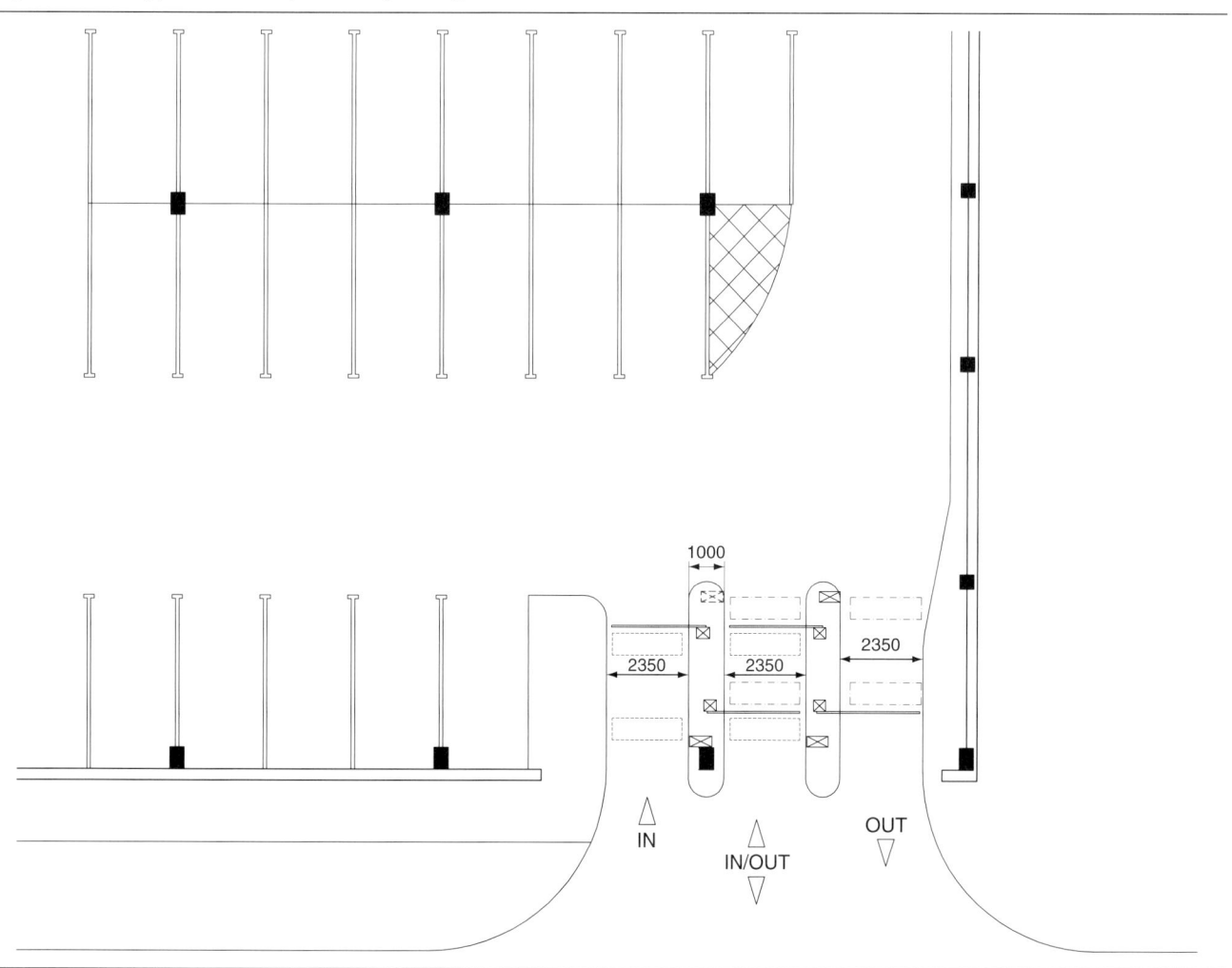

Lane widths at barriers should not be wider than 2.35 m to avoid stand-off situations where motorists are unable to reach the ticket 'spitter' or card reader without leaving their vehicle. Left-turns onto barriers should be avoided for the same reason unless the control barrier is set back from the turn. To avoid roll-back situations, gradients at barriers should be kept as flat as possible and preferably no steeper than one in 30 (3.3%).

The number of barriers to be provided at the entrance and exit will depend on the expected peak-hour traffic flows. The typical working capacity of a single ticket-operated barrier will be approximately 350 vehicles per hour. If a Tag system were to be adopted, the figures per barrier will rise, possibly up to 1000+ vehicles per hour.

In some town centre sites, where space is limited, a four-barrier requirement can be reduced to three by making the middle lane dual-directional, incorporating a barrier at each end, one of which is raised, dependent on the intensity of the traffic flow in a particular direction (see Figure 17.3). Where space is available, an additional barrier can be provided to eliminate queuing problems caused by mechanical breakdowns that occur from time to time.

The entry and exit locations, often referred as entry plazas are a key feature to elevate the car parks status and appeal, and should be easily navigable, well lit and signed, and designed to deal with peak-flow traffic in an efficient manner. The area has a lot of vehicle movements, with drivers giving attention to the exit process, and so pedestrian movement should be restricted, with defined walkways and local barriers to maintain separation. To cater for ticket mis-reads or forgotten tickets, an adjacent pull-off space is a means to keep exit traffic flowing while the problem is dealt with. Where unmanned, the barrier ticket-collector posts should have an audio link to a help point.

Car Park Designers' Handbook
ISBN 978-0-7277-5814-9

ICE Publishing: All rights reserved
http://dx.doi.org/10.1680/cpdh.58149.163

Chapter 18
Ventilation

18.1. Discussion

Insufficient attention to ventilation problems can lead to a build-up of noxious fumes within the car park that can cause nausea and result in the reduced use of the building by the parking public. Where car parks are located underground, mechanical means of ventilation are almost unavoidable, but when they are above ground it is possible to ventilate them naturally or with mechanical assistance. Ventilation requirements are not the result of precise analysis, but have developed empirically through past experience. They are based on common sense and require a modicum of common sense to be used if, while satisfying the requirements overall, local pockets of still air are allowed to occur in which exhaust gases can build up.

Air is 'lazy' and must be directed if it is to be effective. It is of little use locating an extract fan next to an open window. Where natural ventilation is employed in a long rectangular building, ventilating the two flank walls will be quite effective. Where the building is square on plan, distributing the required opening areas evenly around the perimeter can be equally effective. For non-rectangular shapes a common-sense approach to the requirements should be adopted.

Approved Documents B and F of the Building Regulations (NBS, 2010b, 2010c) provide guidance for compliance with the various conditions. BS 7346 (BSI, 2006b) provides further guidance in respect of assisted natural ventilation and mechanical ventilation.

18.2. Natural ventilation requirements

The basic tenets of natural ventilation are as follows.

- It is provided by openings in the elevations at each floor level. The total area of these openings should be at least 5% of the floor area. Usually, when two sides only are ventilated, the required open area is divided equally on each side, but it is allowable for one side to have as little as 1.25% of the floor area, in which case the other side must increase to 3.75% in order to conform with the 5% requirement. Proximity to boundary conditions that reduce the allowable open area on the boundary side can benefit from this rule.
- Two opposite sides can each incorporate 1.25% openings, but the remaining 2.5% must be located somewhere on the other two sides.
- The openings should be spread evenly along the wall lengths, and in order to prevent the local build-up of fumes within the parking areas, extended lengths of solid wall construction should be avoided, especially if it results in right-angled corners. However, if this cannot be avoided, then a soffit-mounted fan can be used to move the stagnant air but still maintain the principle of an open-sided car park, provided that the openings still equate to 5% of the floor area.

18.3. Mechanically assisted natural ventilation requirements

When the opening areas total less than 5%, but more than 2.5% of the floor area, a mechanical ventilation method can be employed to boost the movement of air. It should be capable of providing three air changes per hour, at least. Smoke vents at ceiling level may be also be used when assessing the available open area (see Figures 18.1 and 18.2).

18.4. Mechanical ventilation requirements

Fully enclosed basement and above-ground car parks must employ mechanical ventilation systems that achieve minimum operating standards as follows.

Figure 18.1 Exhaust fans in basement car park

- Six air changes per hour throughout the car park, increasing to ten air changes per hour at exits, ramps and where vehicles are likely to be queuing with their engines running.
- Limiting the concentration of carbon monoxide particles to no more than 50 parts per million averaged out over an eight-hour period and 100 parts per million for peak concentrations at ramps and enclosed exits (this system also needs to be operated at ten air changes per hour throughout the car park during fire conditions. Occasionally, this last requirement is relaxed when it forms part of an engineered approach as discussed below).

Figure 18.2 Induction fan used to boost movement of air

Until recently, mechanical ventilation systems have adopted the use of ducting to carry the air around, usually soffit-mounted and incorporating inlet and outlet grilles. These systems tend to be visually intrusive and expensive and are losing favour compared with the most recent approach of using soffit-mounted jet or impulse fans. The systems are based on ventilation principles developed from tried-and-tested tunnel ventilation techniques, and they differ from traditional systems as follows.

- Ducting is not used: soffit mounted fans control the airflow both at floor and ceiling height.
- Smoke management and control, not usually possible with ducts, is a key feature of the system.
- In larger car parks, the engineering design is based on a design fire size rather than the simple air change requirements referred to in the Building Regulations.

The basic principle is that main extract fans provide the air change rate within the car park while the soffit-mounted fans control the air direction. The internal environmental conditions are constantly monitored by the use of carbon monoxide and smoke-control detectors. The system can be designed to such an extent that deck areas near a fire can be kept free of smoke, thereby aiding access for fire-fighting personnel.

Car Park Designers' Handbook
ISBN 978-0-7277-5814-9

ICE Publishing: All rights reserved
http://dx.doi.org/10.1680/cpdh.58149.167

Chapter 19
Structure

19.1. Discussion

A modern multi-storey car park would be built using a modular form of construction designed with column locations and spans to suit the standard geometric layout for the driving aisles and parking bays. Clear spans would typically be up to 16 m with columns located at one, two, three or four parking bay centres. Providing clear-span construction will produce an open, user-friendly feeling, and locating the columns at the back of parking bays will not only improve the dynamic traffic-flow efficiency of the car park but also reduce the risk of impact damage to parking vehicles.

Floor-to-floor heights will vary depending on the clearance required and depth of structure. A minimum clearance of 2.15 m will allow for most vehicles on the road to park in the facility and by using a depth structure of 450 mm plus an allowance for deflection, it is possible to achieve minimum floor-to-floor heights of 2.65 m even with clear-span construction.

However, additional clearance is usually required for lights and signage and some structural zones can be up to 900 mm, thereby increasing the floor-to-floor height to 3.1 m, allowing for deflection. Beyond 3.1 m is possible but it can present problems with ramp gradients, especially with internal-ramp, split-level-type car parks.

Cost, aesthetics and long-term durability will be factors in the choice of construction. Clear-span flat-soffit concrete-deck solutions can give a more aesthetically pleasing internal environment than the more functional utilitarian construction associated with steel-framed structures, although with proper consideration to design detail, steel-framed car parks can look elegant (see Figure 19.1).

Figure 19.1 Clear-span structure

19.2. Design criteria
19.2.1 Loadings

Imposed loadings for car parks are usually based on the facility being limited to the parking of cars, light vans, and so on, not exceeding 2500 kg gross mass. The imposed loading required for decks and ramps is 2.5 kN (kilonewtons) per m^2. Uniformly distributed load or 9 kN concentrated point load should be applied in accordance with BS EN 1991 (BSI, 2006a) or BS 6399 Part 1. Since this loading is fairly conservative when you consider the actual loads imposed from parked vehicles, there is usually no need to add for additional loads relating to services. Also, it is not usually required to allow for additional snow loads to top exposed decks.

Dead loads will vary depending on the type of structure used and this can vary considerably, typically from 2.7 kN/m^2 to 7.2 kN/m^2. With such a variation available then clearly the choice of structure will be significant when considering the design of the foundations, especially when looking at facilities with numerous floors.

19.3. Stability

In addition to the vertical loads, the facility will require measures to resist horizontal loads due to wind, and the horizontal component of vertical loads applied with a notional inclination. These loads are covered by BS EN 1991 (BSI, 2006a). Resisting these loads is achieved by means of providing vertical bracing, in the case of steel structures, and shear walls, and by transferring loads to stair/lift cores through the diaphragm action of the floor slab. In relation to the location of shear walls, these should not be such that they will affect inter-visibility between drivers and pedestrians nor should they be located in positions that would obscure CCTV camera surveillance required for security purposes.

19.4. Robustness

The current Building Regulations require buildings to be constructed so that in the event of an accident the building will not suffer collapse to an extent disproportionate to the cause. Approved Document A (NBS, 2010a) uses a risk-based approach to classify buildings according to their height and use. Car parks will fall into Building Class 2B for a facility exceeding 6 storeys and into Class 3 for a taller facility. Robustness is achieved by the provision of horizontal and vertical ties through columns and floor plates designed in accordance with the relevant design code for the structure.

19.5. Edge protection

There are three basic considerations for the provision of edge protection.

1. Containment of an errant vehicle within the structure.
2. To prevent a pedestrian from accidentally falling.
3. To discourage people jumping from elevated decks to commit suicide.

The requirements for the design of the edge protection referred to in the first two items listed above are covered by the Building Regulations BS EN 1991-1-7 (BSI, 2006a) and BS 6108-1. However, there is no such guidance with respect to the third item in the list and the designer will have to take a view on this aspect and discuss with the client the risks associated with such acts.

In choosing an appropriate Vehicle Restraint System, it is important to consider the likely deflection of the barrier under impact load. Some barriers rely on adequate space, up to 300 mm, for deflection, in which case adequate space will be required on the inside face of any cladding in order to avoid risk of damage or dislodging the cladding.

19.6. Fire protection

Most open-sided, free-standing car park structures can be constructed in steelwork or concrete without the need for additional fire protection measures since the minimum period of fire resistance required for the structure up to 18 m in height is only 15 minutes. However, for enclosed car parks and basement car parks the fire resistance requirements can increase up to 90 minutes in which case fire protection measures will be required.

Escape stair enclosures will require 60 minutes of fire protection with a 30-minute fire door unless they are fire-fighting shafts containing a fire main, in which case 120 minutes of fire resistance with 60-minute fire doors will be required.

19.7. Vibration

The public's perception of vibration in the decks of car parks is not likely to create alarm as it would in other buildings. A balance has to be struck between the desire to achieve clear-span structures (typically 16 m spans) against economy of build. With the heavier concrete structures, achieving a satisfactory dynamic response it is not so difficult, due to the damping effect of the self-weight. However, with a lightweight steel framed structure, the industry-designed standard is to design for a natural frequency above 3 Hz.

19.8. Durability

Car parks, by their very nature, are often exposed to a harsh environment, compared with other forms of building structures, which usually become enclosed with cladding and finishes. The biggest problem is contamination from de-icing salts that are indirectly deposited on decks from vehicles during the winter months. These salts, when dissolved in water, can percolate through cracks in the decks and cause corrosion of reinforcement leading to spalling of the concrete.

Air quality will also be a factor to be considered when determining the cover required for reinforced concrete elements, and the protection requirements for steel-framed structures.

The durability of any protection systems will be influenced by the corrosivity of the environment. Corrosivity categories can be found in ISO 12944 Part 2 (ISO, 1998) and ISO 9223 (ISO, 2012).

Some basic rules to ensure long term durability are as follows.

1. Determine the working-life requirements and exposure conditions, and choose appropriate concrete mixes with cover to reinforcement from BS 8500-1 (BSI, 2012). Note: the exposure class XD3 should be used on ground-floor or entry decks. Other decks can be considered as exposure class XD1.
2. Ensure all decks have adequate cross-falls (of at least one in 60) leading to adequate drainage points. Note that on no account should open-sided car parks be constructed with flat decks.
   Windblown rain will collect and stand on decks in areas where the structure has deflected. Likewise, do not locate gullies next to internal support columns; they are likely to be left high and dry with ponding water forming at mid-span between the supports.
3. Shrinkage control: ensure that there are adequate movement joints/construction joints to control shrinkage and thermal movement. Most contractors prefer to place concrete in large pours; this is difficult to control. In this case, saw-cut crack-inducing sealed joints should be introduced as soon as the concrete deck is able to take the weight of workmen and saw-cutting equipment.
4. Coatings: consider the application of sacrificial coatings to all decks. Top decks will require a full waterproofing layer using one of the many types of lightweight elastomeric systems in the market place or a mastic asphalt system. Intermediate decks are now often coated using an epoxy applied system for aesthetic reasons, however such coatings will also add to the long-term durability of the structure.
5. Maintenance regime: any new facility should be accompanied with an inspection and maintenance regime. More details on this subject are contained in the Institution of Civil Engineers publication *Recommendations for the Inspection, Maintenance and Management of Car Park Structures* (ICE, 2012).

19.9. Common forms of structure

The following list items identify various forms of structuring currently being used for the construction of car parks. These examples are just a few of the many alternatives available and it is not the intention of this chapter to cover all variants and other solutions available.

1. Steel-framed structure with precast hollow core units spanning onto a steel frame located at 7.2 m–7.5 m centres. Precast concrete floor units would be compositely connected to steel beams by a reinforced concrete topping and offsite welded shear studs to the top of the beams. The corrosion resistance for the structural steelwork elements for this solution would be required to be considered in detail. Typically, the steelwork would be galvanised to 85 or 140 microns thick since the top flange of the

Figure 19.2 Steel frame with hollow core units spanning 7.2–7.5 m (reproduced courtesy of Ramboll UK Ltd)

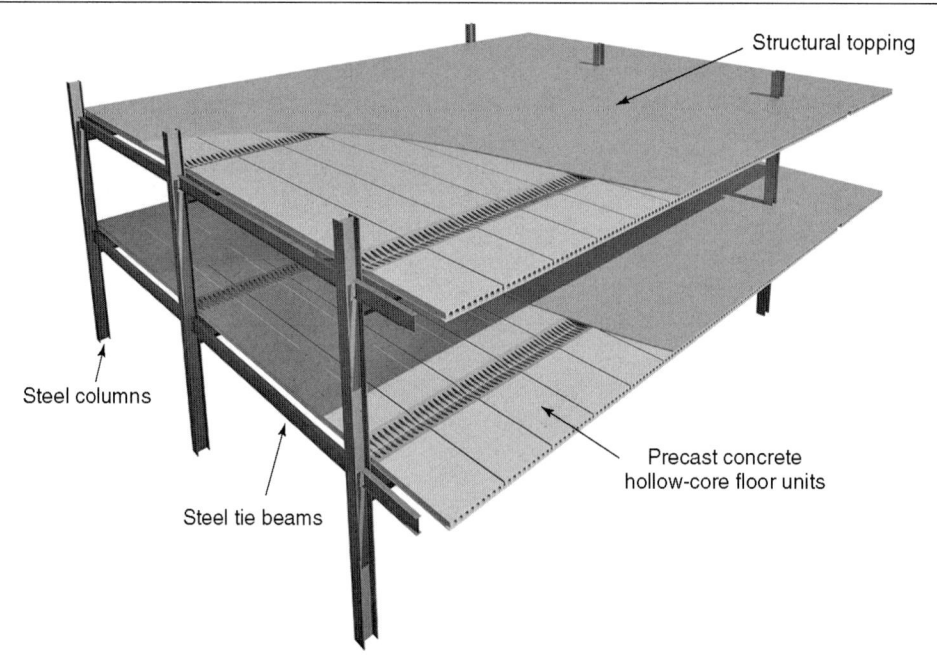

beam will not be exposed for future maintenance typically required at regular periods for a paint-applied coating (see Figure 19.2).

2   Steel-framed structure with composite profiled metal deck/concrete slab (see Figure 19.3). In-situ concrete slab (mesh or fibre reinforced) on profiled metal decking suspended on primary and secondary steel beams and structural steel columns. Concrete slab compositely connected to steel beams by through-deck welded shear studs. The corrosion resistance for the structural steelwork elements and the profiled metal deck for this solution would need to be considered in detail (see Figure 19.4).

3   Pre-cast concrete structure with double-tee concrete floor units. Pre-cast concrete double-tee units spanning 15.7 m onto profiled pre-cast concrete perimeter beams on pre-cast concrete columns. In situ concrete topping required to top surface of slab (see Figure 19.5).

4   Post-tensioned concrete structure with band beams. Post-tensioned concrete floor slabs spanning 7.2–7.5 m onto post-tensioned concrete band beams spanning 15.7 m onto precast or in situ concrete columns (see Figure 19.6).

Figure 19.3 Steel frame with composite profiled metal and concrete deck (reproduced courtesy of Ramboll UK Ltd)

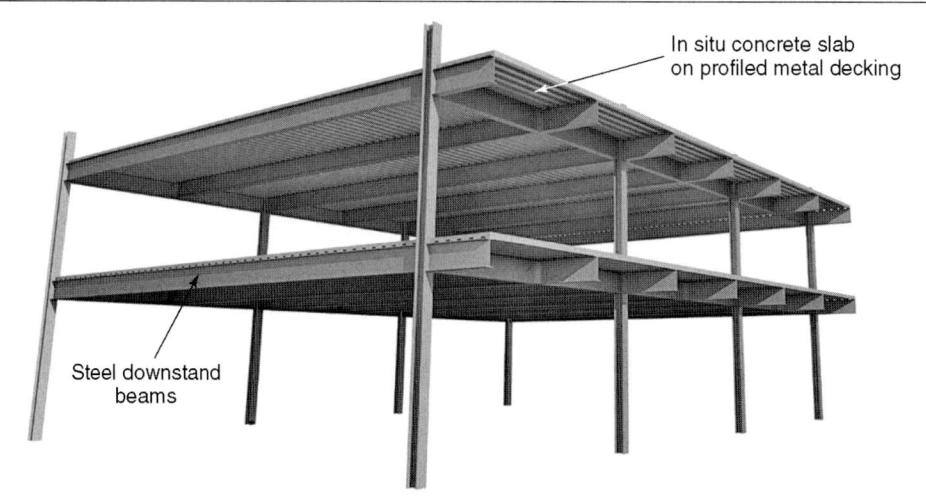

Figure 19.4 Clear-span structure using precast concrete double-tee units (reproduced courtesy of Ramboll UK Ltd)

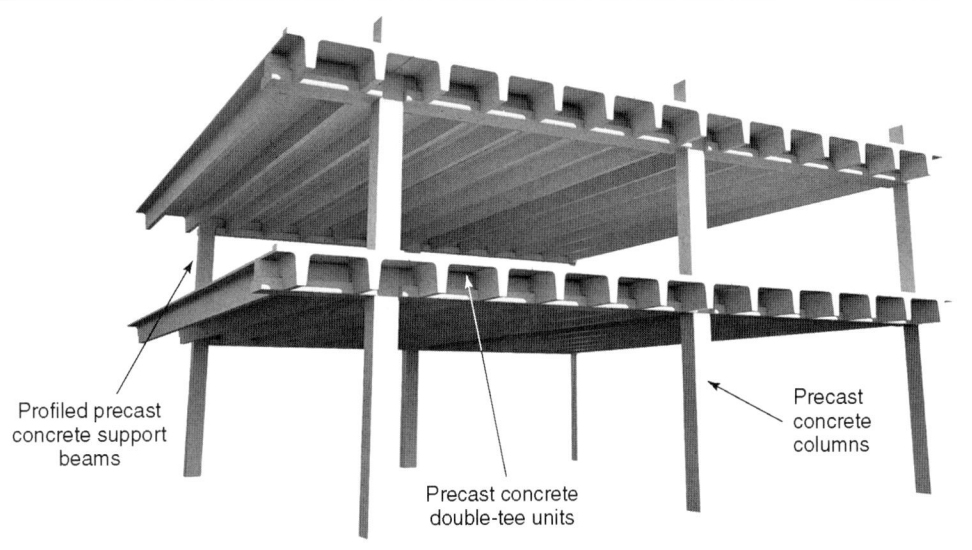

Figure 19.5 Post-tensioned concrete flat slab with drop-down beams (reproduced courtesy of Ramboll UK Ltd)

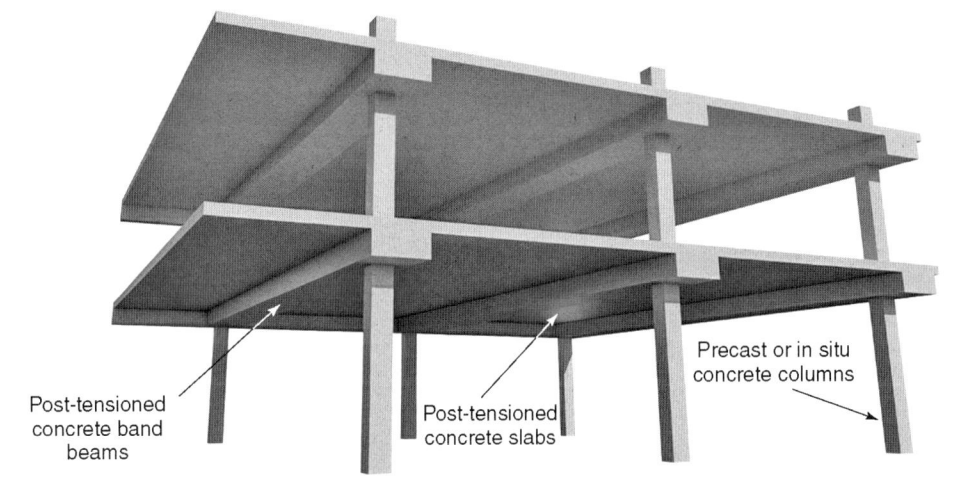

Figure 19.6 Clear-span hollow-core units supported on longitudinal spine beams (reproduced courtesy of Ramboll UK Ltd)

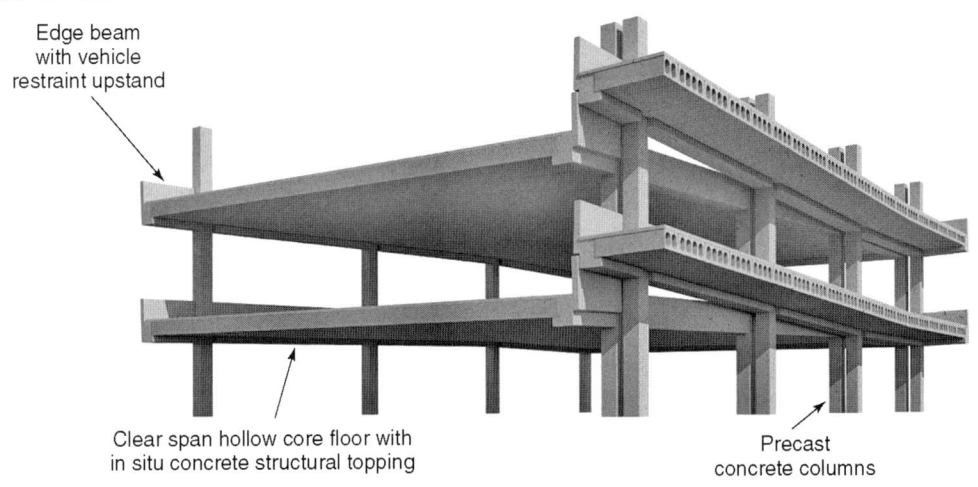

5   Clear-span hollow-core units, spanning 15.6 m with in situ concrete structural topping supported on longitudinal precast spine beams and columns. Hollow cores are broken out at the supports to create a reinforced in situ concrete tie between the slabs and spine beams. Spine beams often incorporate an upstand wall used for vehicle impact resistance.

Car Park Designers' Handbook
ISBN 978-0-7277-5814-9

ICE Publishing: All rights reserved
http://dx.doi.org/10.1680/cpdh.58149.173

Chapter 20
Appearance

20.1. Discussion

Appearance is not just about the exterior envelope. It is essential to the success of any car park that it should provide a safe and attractive internal environment. Before the design of the external envelope is fixed too firmly, consideration should be given to optimising the available site area and creating an efficient vehicle circulation layout. Allowing purely architectural considerations to determine the interior design can often affect, adversely, dynamic and static efficiency and increase the cost per car space beyond that which the client is prepared to accept (see Figure 20.1).

Multi-storey car parks are, basically, utilitarian buildings not too dissimilar from large warehouses and in both cases it is, usually, the aim of the client to gain maximum efficiency of use and economy of construction. The main difference between them is their relative locations. Warehouses occur mainly in commercial districts, docksides and other transportation terminals while car parks occur primarily in urban centres and other highly visual locations.

For both, their architecture comprises mainly of an envelope containing a functional interior. In the case of car parks the function is that of containing cars efficiently and effectively and providing an adequate environment for pedestrians. In town centres, their massing and shear bulk create unique visual problems, but also offer opportunities for interesting sculptural compositions (see Figure 20.2).

The economic construction of a car park structure usually involves some form of modular construction on a fixed grid system. In the case of a clear span structure it will usually be 15.6 m (for one-way flow) or 16.5 m (for two-way flow) in one direction and multiples of the car bay widths in the other. An elevation treatment that allows flexibility in the

Figure 20.1 Eastside Car Park, Birmingham

Figure 20.2 Clarence Dock Car Park, Leeds

choice of these grid configurations will, in the case of 'Design and Build' projects, provide contractors with maximum choice when considering the most appropriate structural framework for a sub-contract package. It should be appreciated that modules of two or three bay widths can produce the most economical solutions for frameworks. Clear-span pre-stressed ribbed-deck systems can also provide economical solutions.

When terracing a car park elevation and where parallel to the structural grid it should occur in modules of parking bay widths or, preferably, on grid lines. Where located at right angles to the structural grid, terracing will require columns to support the new deck edges. In most cases, the columns will need to be carried down through the car park where they will have an adverse effect on parking efficiency and user appreciation.

Car parks nowadays are bought and sold in increasing numbers and market values depend more on their popularity with the parking public than on their external appearance. That does not mean to say that appearance does not matter but that it should not detract, too much, from the car park's dynamic and static efficiency (see Figure 20.3).

20.2. Appearance requirements

Many local authorities have their own requirements relating to the appearance of a car park. Typically these could include the following.

- Height, scale, massing and choice of materials is to be sympathetic with the adjoining buildings and environmental requirements.
- Stair and lift towers should be expressed by height and form to identify pedestrian entry and exit locations.
- Where blank walls are unavoidable, elevation relief features and decoration are advisable.
- The appearance and security aspect of ventilation openings can be enhanced by the use of decorative grills and/or louvres.
- Glazing features should be added to stair and lift cores to aid lighting levels and promote personal security.
- Parked cars should be screened from external view, but views out of the car park from within should be promoted.
- Building forms and elevations at street level should be to a human scale and incorporate weather protection, canopies and low-level landscaping.
- Lightweight roof structures over the top deck or the inclusion of false mansards are sometimes required to screen parked vehicles, especially when adjoining buildings overlook the car park.

In city and town centre sites, it is becoming increasingly common for local authorities to require commercial or retail areas to be included at street level but where they intrude

Figure 20.3 Lighting rail feature to top deck

into the body of the car park they can have an adverse affect on vehicle circulation routes and natural ventilation. However, with careful creative planning from the outset, such problems can be overcome (see Figure 20.4).

Figure 20.4 Q-Park Charles Street car park in Sheffield, nicknamed 'The Cheese Grater'

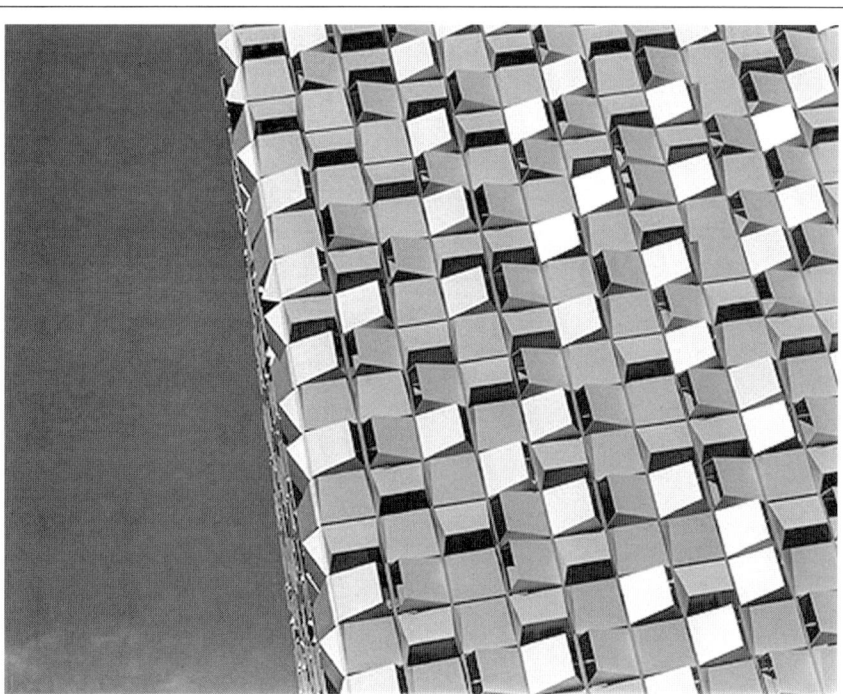

Car Park Designers' Handbook
ISBN 978-0-7277-5814-9

ICE Publishing: All rights reserved
http://dx.doi.org/10.1680/cpdh.58149.177

Chapter 21
Appendix A

A selection of new cars registered in the UK (2013)

		Length: mm	Width: mm	Height: mm	Sold	Market
Mini	Hyundai i10	3565	1595	1540	23 135	35.70%
Super Mini	Ford Fiesta Zetec 1.0 5d	3969	1978	1468	109 265	14.70%
Lower Medium	Ford Focus 1.6 Studio 5d	4358	2010	1461	83 115	16.40%
Upper Medium	BMW 3 Series	4624	2031	1416	44 521	18.70%
Executive	Mercedes-Benz C-Class C180 4d	4591	2008	1447	37 261	31.80%
Specialist Sports	Mercedes-Benz SLK	4134	2006	1301	6668	14.50%
Dual Purpose	Range Rover Evoque 2.2e D4 5d	4365	2125	1635	18 143	9.00%
MPV	Vauxhall Zafira 1.6i 5d	4467	2026	1635	18 401	15.60%

Sources:
http://www.Parkers.co.uk
SMMT (2013)

Chapter 22
References

BPA (British Parking Association) (1970) *Technical Note #1: Metric Dimensions for Car Parks – 90° Parking*. BPA, Haywards Heath.

BPA (1980) *Multi-storey Car Parks in Shopping Centres and Office Blocks*. Report of a seminar in October. BPA, Haywards Heath.

BPA (2013) *Park Mark – The Safer Parking Scheme*. http://www.britishparking.co.uk (accessed 28/07/2013).

BRE (Building Research Establishment Ltd) (2005) *ODPM Building Regulations Division Project Report: Effect of Local Acts on Fire Risks*. BRE, Watford.

BSI (1985) BS 8301:1985: Code of practice for building drainage. BSI, London.

BSI (1989) BS 6571-4:1989: Vehicle parking control equipment. Specification for barrier-type parking control equipment. BSI, London.

BSI (2000) EN 12056-3:2000: Gravity drainage systems inside buildings. Roof drainage, layout and calculation. BSI, London.

BSI (2003) BS EN 13201-2:2003: Road lighting Part 2. BSI, London.

BSI (2006a) BS EN 1991-1-7:2006: Eurocode 1. Actions on structures. General actions. Accidental actions. BSI, London.

BSI (2006b) BS 7346-7:2006: Components for smoke and heat control systems. Code of practice on functional recommendations and calculation methods for smoke and heat control systems for covered car parks. BSI, London.

BSI (2008a) BS 9999:2008: Code of practice for fire safety in the design, management and use of buildings. BSI, London.

BSI (2008b) BS EN 752:2008: Drain and sewer systems outside buildings. BSI, London.

BSI (2010) BS 8300:2009+A1:2010. Design of buildings and their approaches to meet the needs of disabled people: code of practice. BSI, London.

BSI (2011a) BS 6180:2011: Barriers in and about buildings. Code of practice. BSI, London.

BSI (2011b) BS 5266-1:2011: Emergency lighting. Code of practice for emergency escape lighting of premises. BSI, London.

BSI (2012) BS 8500-1:2006+A1:2012: Concrete. Complementary British Standard to BS EN 206-1. Method of specifying and guidance for the specifier. BSI, London.

Ellson PB (1969) *Parking: Dynamic Capacities of Car Parks*, Report LR 221. Road Research Laboratory, Crowthorne.

HMSO (2013) *The Highway Code*. HMSO, London.

ICE (2011) *Design Recommendations for Underground and Multi-storey Car Parks*. 4th edn. ICE, London.

ICE (2012) *Recommendations for the Inspection, Maintenance and Management of Car Park Structures*. ICE, London.

ISO (International Organization for Standardization) (1998) ISO 12944-2:1998: Paints and varnishes – corrosion protection of steel structures by protective paint systems. Part 2: Classification of environments. ISO, Geneva.

ISO (2012) ISO 9223:2012: Corrosion of metals and alloys – Corrosivity of atmospheres – Classification, determination and estimation. ISO, Geneva.

NBS (National Building Specification) (2010a) *Approved Document A – Structure (2004 Edition incorporating 2010 amendments)*. NBS, Newcastle.

NBS (2010b) *Approved Document B (Fire safety) – Volume 2 – Buildings other than dwellinghouses (2006 Edition, incorporating 2007 and further 2010 amendments)*. NBS, Newcastle.

NBS (2010c) *Approved Document F – Ventilation (2010 edition) Approved Document F – Ventilation (2010 edition)*. NBS, Newcastle.

SMMT (Society of Motor Manufacturers and Traders Ltd) (2013) *Motor Industry Facts, 2013 – New Car Registrations by Segment*. SMMT, London.

Car Park Designers' Handbook
ISBN 978-0-7277-5814-9

ICE Publishing: All rights reserved
http://dx.doi.org/10.1680/cpdh.58149.181

Index

4WD (four wheel drive) vehicles 7
45° parking 10–12, 31–32, 40–41, 149
50° parking 31–32
60° parking 12, 31–32, 41
70° parking 11–12, 31–32
75° parking 12, 40–41
80° parking 12, 31–32
90° parking 10–12, 19, 31–32, 40–41, 149
180° turns 8

access
   cyclists 133
   disabled parking stalls 131
   motorcycles 134–6
access assistants 9, 129–31
accessibility, gradients 14
access-ways 13–22, 28–9
ADC (actual dynamic capacity) 29–30
advice signs 149
air changes 163–5
airports 126
aisles
   alignment 9
   dynamic capacities 29–30
   luminance recommendations 146
   pedestrians 147–9
   ramp projections 15–16
   signage 147–9
   widths 10–13
alignment, aisles 9
anchor stands 133, 135
angled parking
   circulation layout 40–1
   design elements 10–12
   dynamic efficiency 30
   ramp widths 19–20
   relative efficiencies 31–2
   see also individual parking angles . . .
ANPR (Automatic Number Plate Recognition) 139, 149, 161
appearance 173–5
Approved Document B 121–7
area per car space 32
arrows 147–9
assistants to disabled drivers 9, 129–31
attenuation tanks 152
automated car parks 142–4

barrier control 159, 161–2
bicycles 133–6
bins 12, 31–2
   see also aisles, stalls

blind corners 138
block zoning 149
BS 8300 2009 regulations 130
building regulations, *Approved Document B* 121–7

capacity
   barrier systems 162
   circular sloping decks 104
   circulation efficiency 32–3
   lifts 124–7
   ramps 111
   recommendations 32–3
   retail estimations 125
carbon monoxide concentrations 164
cast-in-slot drains 153
categories of use 9–10
   see also long-stay car parks; medium-stay car parks; short-stay car parks; tidal car parks
category 1 usage *see* short-stay car parks
category 2 usage *see* medium-stay car parks
category 3 usage *see* long-stay car parks
category 4 usage *see* tidal car parks
CCTV 137, 138–9, 149
The 'Cheese Grater' 175
children 148
circular ramps 87–8, 112–15, 120
   design elements 21–2
   gradients 14–15
   interlocking 22, 120
   super-elevation 24
circular sloping decks (CSD) 103–5, 141–2
circulation design 35–8
circulation direction 40
   see also circulation layouts; one-way flow, two-way flow
circulation efficiency 17–22, 32–3, 36–7
circulation layouts 39–120
   circular sloping decks 103–5, 141–2
   design 35–8
   external ramps 111–20
   flat parking decks with storey height internal ramps 84–90
   flat and sloping decks 67–9, 79–80
   half-external ramps 106–10
   minimum dimension designs 91–102
   parking angles 40–1
   sloping parking decks 53–66
   split-level decks 41–52
   user-friendly features 39–40
   vertical circulation modules 70–8
   warped parking decks 70, 81–3
Clarence Dock, Leeds 174
clean air 163–165

181

Index

clearance 14, 24, 145, 167
clear-span construction 137, 167
clear-span hollow-core units 171–2
client design briefs 5
common structural forms 169–172
composite profiled metal deck/concrete slab construction 170–1
congestion, turn-over rates 29–30
control signs 149
control systems 159–61
cross-overs 40
cross-ramps 13–22, 29
CSD 1 (circular sloping decks) 103–5, 141–2
culs de sacs 40, 78, 84

dead ends 40, 78, 84
dead loads 168
decks
    air changes 163–4
    angled parking efficiency 11
    angled stall area usage 10–12
    area usage and parking angles 10–12
    dynamic capacities 29–30
    gradients 24
    markings 147–9
    roofs 151–2
    see also circulation layouts
de-icing salts 169
Department for Transport (DfT) compliant disabled bay layout 131
design, circulation 35–8
design briefs 5–6
design elements 7–25
    aisle widths 10–13
    angled parking 10–12
    bin dimensions 12
    common structural forms 169–72
    gradients 14–15, 22–4
    height and headroom 24–5
    kerbs 22–4
    lighting 145–6
    parking categories 9–10
    ramps and access-ways 13–22
    robotic parking 142–4
    security 137–40
    stalls 9–12
    standard design vehicle 7–8
    underground parking 141–2
design flexibility 2–3
Des. Recs. (Design Recommendations for multi-storey and underground car parks) 1
dimensions
    bicycle stands 135
    circulation layouts 39
    disabled parking stalls 130–1
    lifts 124–7
    motorcycle spaces 133–4
    new cars sold in 2013 177
    signage 150
    single car lifts 144
    stalls 9
directional signs 147–9
direction of circulation 40
disabled drivers 129–31
    escape routes 124

gradients 14
    luminance recommendations 146
    recommended stall widths 9
doors 123–4, 126, 130, 139
double tees 170–1
double trolley stacking 125
drainage 15, 151–3
drive-through vehicle lifts 143
dry risers 124, 157
ducting 165
durability 169
dynamic capacity
    cross-ramps 29
    exit, entry and internal flow rates 27–8
    parking decks 29–30
    ramps and access-ways 28–9
    stall widths 28
dynamic considerations 27–30
dynamic efficiency 30
    angled parking 30
    rain effects 27
    ramps 13–14
    ramp widths 19

Eastside, Birmingham 173
economic construction 173–174
edge protection 168
edge ramps 89–90
efficiency *see* circulation efficiency; dynamic efficiency; static efficiency
egress *see* exit points
electronic tags 161
elements of design 7–25
    see also design elements
emergency exits 149–50
emergency lighting 146
emergency signs 149–50
emergency stairwells 156
end ramps 56–8, 74–8, 92–8, 101–2
end stall widths 10
energy usage, lighting 145–6
entrance points
    barriers 161–2
    cyclists 133
    disabled drivers 130–1
    luminance recommendations 146
    motorcyclists 136
    ramps 13–14
    security 139–40
entry rates, dynamic capacity 27–8
ER 1 (circular ramp/two-way flow) 112–13
ER 2 (circular ramps/one-way flow) 114–15
ER 3 (three-slope ramps/one-way flow) 116–17
ER 4 (external express ramps) 118
ER 5 (stadium-shaped interlocking ramps) 119
ER 6 (circular interlocking ramps) 120
escape distances 123–5
escape routes 121–5, 149–50, 155–6, 157
evacuations 121–5
excluded rapid outflows 46–8, 147
exit points
    barriers 161–2
    cyclists 133
    disabled drivers 130–1

Index

luminance recommendations 146
motorcyclists 136
ramps 14
security 139–40
exit rates 27–8
exit signs 147–8
exposed roof parking decks, drainage 151–2
express ramps 87–8, 118
external bins 31
external ramps (ER) 32, 111–20
see also ER series

fans 164, 165
feeling of security 137
FIR 1 (one-way flow/full height internal ramps) 37, 85–6
FIR 2 (one-way flow/express ramps) 87–8
FIR 3 (one-way flow/circular ramps) 87–8
FIR 4 (one-way flow/edge ramps) 89–90
fire escapes 121–5, 155–6, 157
fire-fighting measures 157
fire lobbies 123–4
fire protection 168
fire safety 141, 149–150, 155–157
fiscal control 131, 136, 159–61
flash floods 151–2
flat deck drainage falls 15
flat parking decks with storey height internal ramps (FIR) 84–90
see also FIR series
flat and sloping deck layouts (FSD) 67–9
see also FSD series
flexibility of design 2–3
floor-to-floor heights 167
four wheel drive (4WD) vehicles 7
Freyssinet, Eugène 1
FSD 1 (single helix/two-way flow) 68–9
FSD 2 (single helix/two-way flow) 68–9
FSDR 1 (one-way flow with internal ramp) 79–80
full bay surveillance 139

gas build-up 163–165
glazing 139
gradients
accessibility 14
drainage falls 15, 152–3
parking decks 24
pedestrian ramps 22–3
ramps 14–15
ground clearance, standard design vehicle 7
ground investigation reports 5
ground-level enclosure 139–40
guidance, user friendly design 2

half-circular ramps 107–8
half-external ramps (HER) 32, 106–10
see also HER series
half-storey height ramps 16
headroom 14, 24, 145, 167
height-limitation gantries 24–5
heights
floor-to-floor 167
restrictions 24–5
standard design vehicle 7
vehicles 14
helical ramps *see* circular ramps

help points 138–9, 149
HER 1 (half-circular ramps) 107–8
HER 2 (three-slope ramps) 109
HER 3 (SLD-type ramps) 110
high top disabled vehicles 129
highway-approved signage 149
historical perspectives 1–2
horizontal escape 123
horizontal loads 168

impacts, structural resistance 27
impulse fans 165
inclines 14–15
see also gradients
included rapid outflows 43–5
induction fans 164
inflow rates 30
information signs 149
ingress *see* entrance points
inhibited layouts 35
initial briefings 5
interlocking ramps 22, 119–20
intermediate levels 152
internal bins 32
internal movement 27–8
internal ramps 71–3, 79–80, 84–90
internal vertical structure 137

jet fans 165

kerbs 22–4

landings 122–4
layout
barriers 161–2
CCTV installation 138–9
robotic car parks 143
see also circulation layouts; design elements
LED lighting systems 145
left-hand drive vehicles 9
lengths
stalls 9
standard design vehicle 7
levels, number of 35–6
level surveys 5
lifts 121–2, 124–7, 130, 146, 157
lighting 138–9, 145–6
light pollution 145–6
loadings 168
lobbies 123–4, 139
local authority appearance requirements 174–5
lockable security cages 133–4
lockers 136
longitudinal spine beams 171–2
long-stay car parks 9, 32–3, 126
luminance recommendations 146
lux levels 146

manoeuvring envelopes (ME)
historical consistency 1–2
ramp access 17, 20–1
stalls access 17, 19
standard design vehicle 7–8
turning circle templates 20–1, 23

183

Index

market values 174
maximum occupancy 122
maximum pedestrian travel distances in fires 156
MD 1 (one-way flow between end ramps) 92–4
MD 2 (two-way flow/one end ramp) 95–6
MD 3 (ten stalls wide/one end ramp) 97–8
MD 4 (vertical circulation module/eight stalls wide) 99–100
MD 5 (vertical circulation module/eight stalls wide/two end ramps) 101–2
ME *see* manoeuvring envelopes
mechanically assisted natural ventilation 163
mechanical ventilation 157, 163–5
medium-stay car parks 9, 32, 125
minimum dimension layouts (MD) 91–102
   see also MD series
minimum turning diameters 11–12
mobile payments 160
monitoring, CCTV 138–9
motorcycles 133–6
motorists' choice of destination 1
MPV *see* multi-purpose vehicles
multi-purpose vehicles (MPV) 7
multi-span flat deck bin dimensions 12
music 138

natural ventilation 163
Network Rail compliant disabled bay layout 131
new car registrations, UK 177, 2013
night light pollution 145–6
no entry symbols 147
noxious gas build-up 163–5
number of levels 35–6

obstructions between stalls 10
occupancy, fire safety 122–4
one-way flow
   angled parking 10–12, 19–20
   bin dimensions 12
   circular ramps 87–8, 114–15
   double helix 59–61, 65–6
   edge ramps 89–90
   end ramps 74–6, 92–4, 101–2
   excluded rapid outflow 46–8
   express ramps 87–8
   external ramps 111–20
   half-external ramps 106–10
   included rapid outflow 43–5
   internal ramps 41–52, 71–3, 79–80, 84–8
   markings and signs 147
   ramp widths 17–20
   rapid outflows 43–8, 56–8, 87–8
   scissor ramps 49–50
   single helix 56–8
   three-slope ramps 116–17
open-sided car parks, fire risk 155–6
optimum stall widths 9
outflows
   rapid 38, 43–8, 56–8, 87–8, 147
   rates 30
outlet valves 157
overhead signs 147–9

parking angles *see* angled parking; individual parking angles ...

parking categories 2–3, 9–10
   see also long-stay car parks; medium-stay car parks; short-stay car parks; tidal car parks
parking decks *see* decks
parking times 37–8
Park Mark scheme 137–8, 145–6
part M of *Approved Document B* 124
pay-and-display 149, 159
payment by mobile phone 160
payment-on-exit 159
payment-on-foot 149, 160
pay stations 149, 160
peak flows 37, 106, 122
pedestrian ramps 14–15, 22, 124–5
pedestrians
   aisle widths 10, 12
   escape distances 123–4
   luminance recommendations 146
   signage 147–9
   sloping parking decks 53
   split-level deck designs 42–3, 45, 48, 50, 52
   vertical circulation module layouts 70
petrol interceptors 152
pollutants 163–5
post-tensioned concrete structures 170–1
pre-cast concrete structures with double tees 170–1

Q-Park Charles Street, Sheffield 175

rain 27, 151–153
ramps
   access manoeuvring envelopes 17, 20–1
   air changes 163–4
   aisle projections 15–16
   at ends 56–8, 74–8, 92–8, 101–2
   capacity 111
   circular 21–2, 87–8, 112–15, 120
   design elements 13–22
   dynamic capacities 28–9
   dynamic efficiency 13–14, 19
   entrances 13–14
   escape routes 125
   exits 14
   express 87–8, 118
   external 111–20
   gradients 14–15
   half-circular 107–8
   half-external 106–10
   half-storey height 16
   headroom 14
   helical, gradients 14–15
   interlocking 22, 119–20
   internal 71–3, 79–80, 84–90
   kerbs 22–4
   luminance recommendations 146
   one-way flow 17–20
   pedestrian 15, 22, 124–5
   scissor-type 21
   side-by-side 21
   side clearance 17
   stadium-type 22, 119
   storey height 17–18, 84–90
   three-slope 109, 116–17
   transitional slopes 15

two-way flow 20
widths 17–20, 29
rapid inflows 38
rapid outflows 38, 43–8, 56–8, 87–8, 147
reduced aisle widths 12
reflective surfaces 145
refuges 124
relative efficiency 31–2
Report LR221 1
retail lift capacity 125
right-angle parking 10–12, 19, 31–2, 40–1, 149
right-hand drive vehicles 9
ring spanner-shaped layouts 91, 92–4
robotic car parks 142–4
robustness 168
roof decks 146, 151–2
roofs 36

safety
   cycles and motorcycles 133–6
   fires 121–5, 141, 155–7
   signage 149–50
salts for de-icing 169
scissor-type ramps 21
SD 1 (single helix with two-way flow) 54–5
SD 2 (single helix/one-way inflow and rapid outflow) 56–8
SD 3 (double helix/one-way central aisle flow) 59–61
SD 4 (double helix/end connected two-way central flow) 62–4
SD 5 (double helix/one-way flow) 65–6
SDV *see* standard design vehicle
security 137–40
security cages 133–134
security fences 140
Sheffield stands 133, 135
Shopmobility units 131
shortest travel distance 36
short-stay car parks 9, 27, 32, 125
side-by-side ramps 21
side clearance, ramps 17
signage 147–50
simplicity, circulation design 39–40
single bins 31
single car lifts 144
single trolley stacking 125
site surveys 5
sizing, lifts 124–7
SLD 1 (one-way flow/included rapid outflow) 1, 32, 36–7, 43–5
SLD 2 (one-way flow/excluded rapid outflow) 32, 46–8
SLD 3 (one-way flow/scissors-type ramps) 32, 49–50
SLD 4 (two-way flow aisles and ramps) 51–2
sloping parking decks (SD) 53–66
   pedestrians 53
   see also circular sloping decks; SD series
smoke management 165
speed limits 28
split-level decks (SLD) 41–52
   escape routes 124–5
   pedestrians 42–3, 121, 124–5
   see also SLD series
sports utility vehicles (SUV) 7
sprinklers 157
stability 168
stadium-shaped interlocking ramps 119
stadium-type ramps 22, 119

stairs 121–5, 139, 146, 152
stalls
   access manoeuvring envelopes 17, 19
   design elements 9–12
   dimensions 9, 130–1
   disabled drivers 129–31
   luminance recommendations 146
   obstructions 10
   turn-over rates 29–30
   widths 9–12, 28
standard design vehicle (SDV) 7–8
stands for motorcycles 134–5
static capacity 32–3, 130
static efficiency 31–3
   area per car space 32
   capacities 32–3
   circular sloping decks 105
   flat parking decks with storey height internal ramps 86, 88, 90
   flat and sloping decks with internal ramps 80
   half-external ramps 106, 108
   minimum dimension layouts 91, 94, 96, 98, 100, 102
   parking angles 31–2
   sloping parking decks 53, 55, 58, 61, 64, 66
   split-level decks 45, 48, 50, 52
   two-way flow ramps 20
   vertical circulation module layouts 73, 76, 78
   warped parking decks 83
steel-framed structures 169–70
stopping distances 28
storey height ramps 17–18, 84–90
storey heights 24
storm water run-off 151–2
straight ramps, gradients 14–15
structural elements, impact resistance 27
structural forms, common 169–72
structures 167–72
sudden slope changes 15
super-elevation, circular ramps 24
surface car parks 35
surveillance staff 138
SUV (sports utility vehicles) 7
swept path turning circle templates 20–1

tag systems 161
templates, turning dimensions 20–1, 23
terminals 126
three-slope ramps 109, 116–17
tidal car parks 9, 33, 126
time taken parking 37–8
top decks 36, 146, 151–2
transitional slopes 15
travel distances 36, 156
turning circle templates 20–1, 23
turning dimensions 7–8, 11–12, 20–1, 23
turn-over rates, stalls 29–30
two-way flow
   aisles and ramps 51–2
   bin dimensions 12
   circular ramps 95–6, 104–5, 112–13
   double helix 62–4
   dynamic efficiency 27–8
   end ramps 56–8, 77–8, 95–6
   external ramps 112–13
   parking angles 12

Index

two-way flow (*continued*)
   ramps 20–1
   single helix 54–5, 68–9

U-bends 152
underground parking 24, 103, 141–2
uninhibited layouts 35
unmanned barriers 162
upper decks 36, 146, 151–2
user-friendly design 2, 39–40
U-traps 152

Variable message Signs 147, 149
VCM 1 (one-way flow with internal ramps) 32, 71–3
VCM 2 (one-way flow with end ramps) 74–6
VCM 3 (two-way flow with a single end ramp) 77–8
Vehicle Restraint Systems 168
vehicles
   height 14, 24–5
   impacts 27
   left/right-hand drive 9
   speed limits 28
   stopping distances 28
   turning dimensions 7–8, 11–12, 20–1, 23
   see also standard design vehicle
ventilation 163–5
vertical access 121–7
vertical circulation module (VCM) layouts 70–78
   see also VCM series
vertical structures 137, 168
vibration 169
viewing angles 9
visibility 13, 17, 137–9, 145–6
vision panels 139
visually impaired persons 147

walking man signs 147–8
warped parking deck (WPD) layouts 70, 81–83
washing down 153
wheelbases, standard design vehicle 7
widths
   aisles 10–13
   barrier lanes 162
   circular ramps 22
   ramps 17–22, 29
   signage 150
   stairs 122–4
   stalls 9, 28
   standard design vehicle 7
women-only car parks 140
WPD 1 (warped parking decks) 70, 81–3

zoning
   blocks 149
   lighting 146
   motorcycle parking 133–4